Comparative Crystal
Chemistry

Comparative Crystal Chemistry

Temperature, Pressure, Composition and the Variation of Crystal Structure

Robert M. Hazen and **Larry W. Finger**

Carnegie Institution of Washington
Geophysical Laboratory, Washington, DC

A Wiley–Interscience Publication

1807 1982

JOHN WILEY & SONS
Chichester · New York · Brisbane · Toronto · Singapore

Library of Congress Cataloging in Publication Data:

Hazen, Robert M., 1948–
  Comparative crystal chemistry.

  'A Wiley–Interscience publication.'
  Includes index.
  1. Crystallography.   I. Finger, Larry W.
II. Title.
QD921.H435   548   82-2834
ISBN 0 471 10268 7   AACR2

British Library Cataloguing in Publication Data:

Hazen, Robert M.
  Comparative crystal chemistry.
  1. Crystallography
  I. Title   II. Finger, Larry W.
  548   Q905.2

  ISBN 0 471 10268 7

Typeset by Pintail Studios Ltd., Ringwood, Hampshire.
Printed in the United States of America

Dedicated to Roger Strens

Contents

Foreword xi

Preface xiii

1 Introduction 1

PART ONE   EXPERIMENTAL PROCEDURES

2 High-temperature Crystallography 5

   I   Single-crystal Heaters 5
   II Crystal Mounting 12
   III Temperature Calibration 14
   IV Future Prospects 15

3 High-pressure Crystallography 17

   I   Development of the Diamond Pressure Cell 18
   II Diamond-Cell Design 23
   III Operation of the Diamond Cell 33
   IV Future Prospects 54

4 High-temperature, High-pressure Crystallography 57

   I   Heating the Diamond Cell 57
   II Design of a High-temperature Diamond Cell 60
   III Operation of the High-temperature Diamond Cell 64
   IV Future Prospects 75

5 The Parameters of a Crystal Structure 77

   I   Characterizing and Comparing Crystal Structures 77
   II Propagation of Errors 78
   III Changes in the Unit Cell 79
   IV Changes in Atomic Positions 82
   V   Thermal Parameters 86
   VI Conclusion to Part One 88

Appendix I   Suppliers 90

Appendix II   A Program to Calculate the Strain Tensor from
              Two Sets of Unit-cell Parameters 92

Appendix III A Program to Calculate Polyhedral Volumes and
              Polyhedral Distortion Parameters from a Set of Atomic
              Coordinates and a Unit Cell 103

PART TWO   STRUCTURAL VARIATIONS WITH TEMPERATURE,
            PRESSURE AND COMPOSITION

6 Structural Variations with Temperature 115

   I   Introduction to Part Two 115
   II   Effects of Temperature on the Ionic Bond 116
   III Dimensional Variation with Temperature 123
   IV   Thermal Parameter Variation with Temperature 139
   V   Conclusions 142

7 Structural Variations with Pressure 147

   I   Effects of Pressure on the Ionic Bond 147
   II   Dimensional Variations with Pressure 149
   III Other Structural Variations with Pressure 160
   IV   Conclusions 161

8 Structural Variations with Composition 165

   I   Introduction 165
   II   Effects of Composition on Structural Dimensions 166
   III Conclusions 175

9 Continuous Structural Variations with Temperature, Pressure
   and Composition 177

   I   Introduction 177
   II   The Structural Analogy of Temperature, Pressure and
        Composition 178
   III T, P and X variations of α, β and γ 187
   IV   Modelling Structural Variations—Distance Least Squares . . 192

10 Structural Variations and the Prediction of Phase Equilibria . . . 194

    I   Topological Classification of Phase Transitions 194
    II   Geometrical Limits to Phase Stability 196
    III Polyhedral Stability Fields 211
    IV   Conclusions 214

Author Index 217

Subject Index 221

Formula Index 229

Foreword

Over the past 12 years there has been amazing progress in the ability of solid-state scientists to obtain diffraction data from single crystals at high temperatures and pressures. The development of the techniques to do this has required that investigators apply their knowledge of a wide range of material properties and develop experimental skills much advanced over those used in the usual diffraction studies under ambient conditions. R. M. Hazen and L. W. Finger have been pioneers in this field, overcoming obstacles that would have stopped many others. This book is a welcome contribution, for it presents in detail many of the experimental techniques required for successful diffraction experiments under extreme conditions of temperature and pressure. It is not easy to do this work and this book should be of great value to those without previous experience and also to those who are experienced, but want to improve their skills.

I look forward to many more advances in the next 12 years and believe that this book will help very much to accelerate the pace of discovery.

CHARLES T. PREWITT

Preface

An understanding of the crystalline state at nonambient conditions is an integral part of the physical sciences. Materials scientists, solid-state physicists and chemists and solid-earth geoscientists are all routinely confronted with problems involving crystalline matter at extreme conditions of temperature and pressure. New high-temperature and high-pressure crystallographic techniques, developed within the past dozen years, are a direct response to these intriguing experimental problems. To the crystallographer's traditional role of determining periodic atomic arrangements has thus been added the capability of measuring the variation of these atomic arrangements with temperature and pressure.

In spite of the obvious applications of high-temperature and high-pressure crystallography there is to date no comprehensive compilation of data or techniques. This lack is not surprising given the diversity of backgrounds of contributors. Much work has been published by earth scientists concerned with the structures of oxides and silicates at their conditions of formation. Many studies by solid-state chemists and physicists involve the temperature and pressure variation of electronic structure and bonding. Other material scientists are interested in the nature of polymorphic phase transitions. Data on structures at high temperature and pressure are often widely scattered, while information on experimental apparatus and techniques is commonly not even available in the published literature. Perhaps as a consequence of the ever-diminishing 'least-publishable unit', many published descriptions of high-temperature or high-pressure devices are schematic only, or represent early versions of significantly modified equipment. Some widely used apparatus and procedures have never been adequately described in the literature. For example, we know of no detailed published description of the loading of a single-crystal, high-pressure cell. We undertook this book as an attempt to consolidate the diverse literature on the acquisition and analysis of high-temperature and high-pressure crystallographic data.

This monograph has been prepared as an introduction and review of 'comparative crystal chemistry', or the study of crystal structure variations with temperature, pressure and composition. The book is divided into two main sections, which are largely independent. The first, 'Experimental Procedures', is a response to dozens of requests for a step-by-step guide to techniques for the operation of high-temperature and high-pressure single-crystal x-ray devices. Our objective has been to document current experimental methods in sufficient detail that our procedures may be duplicated. Diagrams of experimental apparatus and component specifications of equipment now in use at the Geophysical Laboratory and elsewhere, as well as detailed descriptions of crystal mounting, orientation, calibration and computational procedures, which are not widely available, have been included. Appendices to Part One include a list of suppliers mentioned in the text and program listings for strain tensor and polyhedral volume calculations. The first section is thus designed as a handbook for high-pressure and high-temperature crystallography.

The second part of the book is a summary of the results of several dozen high-temperature and high-pressure crystallographic studies. Although only a small fraction of known structures have been investigated at nonambient conditions, high-temperature or pressure experiments have already yielded insights into the stability and chemical bonding in inorganic compounds. Empirical trends rather than theoretical implications of these structure studies are emphasized, and a simple procedure for predicting structural variation with temperature, pressure and composition is proposed. In several crystalline solids the predictions of structural variations have led to prediction of phase equilibria, and this application of comparative crystallography is also explored.

We have attempted to incorporate pertinent references up to June, 1981. For example, Part Two includes tabulations of all high-temperature and high-pressure, single-crystal, x-ray structure determinations known to us at mid-1981. We would be grateful to receive information regarding published studies that we inadvertently omitted, as well as any new articles on nonambient crystallographic techniques and structure data.

In preparing this monograph we have assumed that readers have a basic knowledge of crystallography, crystal chemistry and single-crystal x-ray diffraction. The work is directed to all solid-state disciplines, and use of specialized terminology (e.g. mineral names) has been avoided. Standard SI units are used throughout, save for retention of the Angstrom unit of distance and the kilobar unit of pressure, which have been cited in virtually all papers in high-pressure and high-temperature crystallography.

* * *

In the summer of 1978 Roger Strens was asked by John Wiley and Sons Ltd. to coordinate and edit a series of monographs on various aspects of mineral physics. The present volume was to have been the first of the project, but Dr. Strens' tragic death in January 1980 brought to an end the proposed series. We gratefully acknowledge Roger Strens' thoughtful advice and enthusiastic support, and we dedicate this volume to Roger, in appreciation for his manifold contributions to mineral physics.

This monograph was completed while both authors were staff members at the Carnegie Institution of Washington's Geophysical Laboratory, which has been the sponsor of many pioneering studies in the behaviour of mineral systems at high temperature and high pressure. It is a special pleasure to thank the

Laboratory and its Director, Hatten S. Yoder, Jr., for providing both generous support and a unique research environment, without which the preparation of this book, as well as many of the studies described herein, would not have been possible.

Many individuals have helped us in the preparation of this book, through conversations, correspondence and the sharing of data and illustrations. In particular, the authors gratefully acknowledge Linda Pinckney and Russell Ralph for their thorough and perceptive reviews of early drafts of this monograph. Charles W. Burnham and Yoshi Ohashi provided detailed and thoughtful analyses of the penultimate manuscript, and their suggestions are reflected throughout the final version of the book. Special thanks are due to Charles T. Prewitt for both his generous advice and his encouragement throughout the preparation of the monograph.

R. M. HAZEN
L. W. FINGER

Chapter 1

Introduction

Von Laue's discovery in 1912 of the diffraction of x-rays by crystals revolutionized the solid state sciences. His observations demonstrated that crystals have a periodic structure with a repeat unit on the order of the wavelength of the x-rays. Methods that used positions and intensities of the diffracted x-rays to infer crystal structure were soon developed. For forty years virtually all x-ray crystallographers applied these techniques to the solution of new crystal structures. These first researchers emphasized determination of unknown atomic topologies, and by the mid-1950s most common inorganic structure types had been solved through their efforts.

As more and more crystal structures were resolved the objectives of many crystallographers gradually changed. Studies of atomic topology were replaced by more detailed investigations of chemical bonding and electron distribution. One common approach to these bonding studies was the determination of the small structural differences between isomorphous compounds. In many isomorphous materials and their solid solutions the subtle variations in atomic parameters within a given topology are related to the differing electronic structures of the substituting atoms. Such *comparative* crystallographic investigations led directly to important concepts of crystal chemistry including the periodicity of ionic radii, the effects of radius ratio on coordination number and geometrical limits of certain atomic topologies (Pauling, 1960).

Another obvious but technologically more difficult problem in comparative crystallography is the measurement of subtle atomic positional and lattice variations which are manifest in thermal expansion and compression. In general, the structural response to changes in temperature or pressure, as with changes in composition, is not a simple scaling of the unit cell. Instead, the size ratios of structural elements change, because of nonuniform expansion or compression of different types of bonds. Time-consuming, complete three-dimensional structure refinements at several temperatures and/or pressures are required to elucidate these changes. The resulting 'structural equations of state' are well worth the effort, however, for they provide important data on both theoretical and applied problems in solid state sciences.

The variation of bond distance with temperature, pressure and composition is closely tied to the shape of the bond potential function, which relates potential energy to interatomic separation. Room-condition refinements of different compositions yield equilibrium separations (in other words the distance at which potential energy is minimized at zero pressure), whereas high-temperature and high-pressure refinements provide information on the asymmetry and depth of the potential well. Comparative studies of the temperature, pressure and compositional variation of a given structure type thus impose constraints on theoretical bonding models.

Empirical descriptions of structural variation with temperature, pressure and composition are of vital importance in predicting the behaviour of materials at conditions not yet studied, or not attainable, in the laboratory. Our empirical knowledge of ionic radii has led to the synthesis of numerous economically valuable analogues of natural compounds. Empirical models of the variation of structure with temperature and pressure may be used to predict material properties such as molar volume or crystal field energies at high temperatures and pressures (for example at conditions in the earth's mantle). Furthermore, if a phase transition is known to be controlled by structural parameters, then phase equilibria may be deduced from the predicted structure at specific temperatures, pressures and compositions.

In spite of the theoretical and empirical significance of comparative crystallographic studies, technological problems of these experiments have been solved only recently. The collection of multiple, precise data sets for comparative purposes was not practical prior to the development of automated single-crystal diffractometer systems. These experimental systems, combined with the arsenal of crystallographic software and high-speed computers, allow the rapid automated completion of tasks that would have required several months two decades ago. Another vital technological advance has been the development of small and stable resistance heaters and diamond-anvil pressure cells capable of maintaining single crystals in a stable orientation at constant high temperature and pressure for the several-day duration of an x-ray experiment. The solution of these formidable technological problems and the application of this technology to basic solid state problems are the subjects of the subsequent chapters of this monograph.

REFERENCE

Pauling, L. (1960) *The Nature of the Chemical Bond and the Structure of Molecules and Crystals*, Cornell University Press, Ithaca, NY. 644 pp.

PART ONE

EXPERIMENTAL PROCEDURES

Chapter 2

High-temperature Crystallography

CONTENTS

I   Single-crystal Heaters 5
    A   Early Developments 5
    B   Types of Heater 6
       (1)   Open-flame heaters 6
       (2)   Gas-flow heaters 7
       (3)   Radiative heaters 8
    C   Design of a Radiative Crystal Heater 8
II   Crystal Mounting 12
III   Temperature Calibration 14
IV   Future Prospects 15

I. SINGLE-CRYSTAL HEATERS

A. Early developments

The determination of crystal structures at high temperatures is of fundamental importance to studies of the solid state. It is not surprising, therefore, that high-temperature devices were applied to x-ray cameras by the 1920s, shortly after the development of powder x-ray diffraction as a useful identification technique. Perhaps the earliest successful high-temperature structure determination was completed at the Geophysical Laboratory, where Wyckoff (1925) studied the cristobalite form of SiO_2 to 430°C. The earliest high-temperature x-ray investigations by Wyckoff and others were on polycrystalline materials. Although many powder heaters were designed in the subsequent decades, there was still no simple and effective single-crystal heater when Goldschmidt (1964) prepared his comprehensive *Bibliography of High-Temperature X-ray Diffraction Techniques*. One of the first practical single-crystal heaters for x-ray diffraction was introduced by Foit and Peacor (1967), who designed their device for Weissenberg geometry. The heater was not easily adapted to automated systems, nor to other diffraction

Table 2-1. Single-crystal heaters for x-ray diffraction studies

Authors	Type	Maximum T (°C)	Stability*	Geometry
Foit and Peacor (1967)	radiative	1000	2	Weissenberg
Smyth (1969)	open flame	1500	100	precession
Prewitt, Papike and Ross (1970)	gas flow	700	15	precession
Viswamitra and Jayalakshmi (1970)	radiative	1000	10	Hilger and Watts linear diffractometer
Lynch and Morosin (1971)	radiative	1000	$\ll 1$	Various
Smyth (1972)	gas flow	900	10	Precession/ 4-circle
Glazer (1972)	radiative	900	<1	Weissenberg
Brown, Sueno and Prewitt (1973)	radiative	1100	15	4-circle
Finger, Hadidiacos and Ohashi (1973)	radiative	900	10	4-circle/ precession
Reeber (1975)	radiative	900	<1	Laue
Ishizawa, Miyata, Minato and Iwai (1978)	gas flow	1000	5	4-circle
Tuinstra and Fraase Storm (1978)	gas flow	700	5	Various
Lissalde, Abrahams and Bernstein (1978)	gas flow	400	<1	Various

* Values of stability are from the literature or from personal communications with users of the device. NB: Some authors quote stability in terms of precision (relative uncertainty in calibration) and others report accuracy (absolute calibration uncertainty). Accuracy is always less than precision in high-temperature calibration.

geometries, and therefore did not receive widespread use. Several other heaters were described shortly thereafter, and by the mid-1970s high-temperature crystal structures were being completed in several laboratories (Prewitt, 1976). A variety of single-crystal heaters for x-ray diffraction are listed in Table 2-1.

B. Types of heater

Three types of single-crystal heater have been employed in crystallographic studies: open-flame, gas-flow and radiative. Each type is described below.

(1) Open-flame heaters

Smyth (1969) employed an open-flame, oxy-hydrogen torch in his experiments on $(Mg_{0.3}Fe_{0.7})SiO_3$, a clinopyroxene-type chain silicate. An obvious advantage of this device is the extremely high temperatures, greater than 1500°C, that may be generated. The open-flame heater is severely limited, however, in that temperature at the crystal is stable to only about ±100°C owing to minor fluctuations in gas flow and air convection. Because of this problem the open-flame heater has received comparatively limited use and will not be considered further here. It should be recognized, however, that open-flame techniques have the potential for significantly higher temperatures than either the gas-flow or radiative designs described below.

(2) Gas-flow heaters

Furnaces developed by Prewitt, Papike and Ross (1970) and Smyth (1972) rely on a hot gas, usually nitrogen, to heat the single crystal. The basic features common to these heaters (Figure 2-1) are a coiled resistance heating element of noble metal wire and a central channel through which high-purity nitrogen gas is directed over the crystal. In addition, the Smyth (1972) heater, which is perhaps the most compact and versatile of the gas-flow designs, has a water-cooled jacket and connecting water lines, which are parallel to the nitrogen gas and power lines.

Gas-flow heaters have proved successful in producing and maintaining stable high temperatures to 900°C for several days, and have thus been employed in more than a dozen high-temperature structure studies. A significant limitation of the technique is that temperatures above 900°C are not attainable. Furthermore, the gas and water lines are cumbersome, and temperature may fluctuate if fluid flow rates are not constant. For these reasons gas-flow crystal heaters have been largely supplanted by radiative designs.

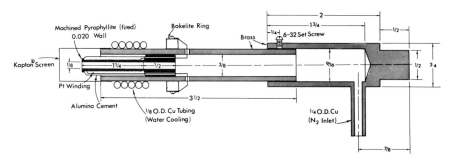

Figure 2-1. Gas-flow heater of Smyth (1972) in longitudinal section.
'A small twist of Pt foil is placed in the cool end of the pyrophyllite tube to cause turbulent flow of nitrogen through the furnace and eliminate temperature gradients across the orifice. The thin wall of the pyrophyllite tube is threaded at 44 turns per inch to separate the windings of 0.015'' diameter Pt wire. A current of 6.0 amperes at 20 volts AC was required to maintain the crystal at 900°C at a distance of 3 mm from the orifice.' (Figure from Smyth, 1972, reproduced by permission of the Mineralogical Society of America)

8

(3)   Radiative heaters

The most versatile single-crystal heaters available today are based on the design developed at the State University of New York at Stony Brook by Brown, Sueno and Prewitt (1973, Figure 2-2). In the original design a miniature resistance heating element in the shape of a horseshoe was placed around the crystal. Temperatures in excess of 1100 °C have been obtained with relatively low power requirements, and no gas or water lines are required. Furthermore, the small size and wide angular access to the crystal facilitate the adaptation of this device to different diffraction geometries. A modification of the Stony Brook design is available commercially (Blake Industries). Another heater, developed by Dr. Yoshikazu Ohashi, is currently in use at the Geophysical Laboratory and is described in detail below.

C.   Design of a radiative crystal heater

The Geophysical Laboratory crystal heater is illustrated in Figure 2-3. Heating elements of the furnace consist of $Pt_{87}Rh_{13}$ thermocouple wire threaded through four-hole thermocouple shield ceramic tubing. These elements are arranged in pairs or triangular clusters on either side of the crystal position. The resistance

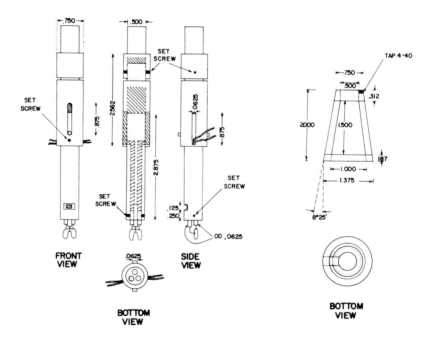

Figure 2-2. Radiative heater of Brown, Sueno and Prewitt (1973) with heater cover.
(Reproduced by permission of the Mineralogical Society of America)

(a)

(b)

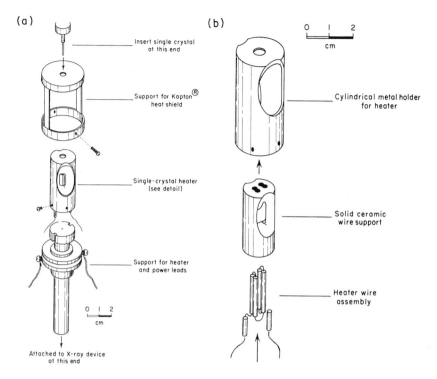

Insert single crystal
at this end

Support for Kapton®
heat shield

0   1   2
cm

Cylindrical metal holder
for heater

Single-crystal heater
(see detail)

Support for heater
and power leads

Solid ceramic
wire support

0   1   2
cm

Heater wire
assembly

Attached to X-ray device
at this end

Figure 2-3. Radiative heater for single-crystal x-ray diffraction, designed by Y. Ohashi. Stippled areas represent ceramics, and all other parts are metal.
A. Exploded view of complete assembly.
B. Exploded view of heater

heater assembly is mounted in a cylinder of machinable ceramic surrounded by stainless steel. Conical openings for incident and diffracted beams are machined, and the entire furnace assembly is supported on the chi-circle opposite the phi-drive system (Figure 2-4). The definition of four-circle angles and coordinate system is given in Chapter 3, Section III-D-1. An external shield of heat-resistant plastic film (Kapton from E. I. Dupont de Nemours and Co., Inc.) is used to reduce convection near the sample.

Temperature control of the Geophysical Laboratory furnace is accomplished by a sensing thermocouple near the crystal position (Figure 2-5) that is used as input to a controller designed by Finger et al. (1973). The e.m.f. of the sensing thermocouple is amplified, displayed on a digital meter, compared with a reference value, and the resulting difference is used to control the output power of a d.c. supply connected to the furnace winding. A novel feature of this system is that the reference voltage may be supplied either from a potentiometer on the front panel or from the output of a digital-to-analogue converter interfaced to the computer

Figure 2-4. Geophysical Laboratory single-crystal heater mounted on the automated four-circle diffractometer

that controls the four-circle diffractometer. The digitized output of the thermocouple amplifier is input to the controller; thus, a means of automatically controlling the temperature is available.

In calibration of the system, differences between the temperatures at the reference thermocouple and the crystal position, as determined from known melting points and output of thermocouples in the sample position, were found to vary significantly and systematically with the angle between the central axis of the furnace and the vertical. Thus, as diffractometer angle chi (χ) changes from 0° (furnace pointing down) to 180° (furnace pointing up), the temperature at the crystal may vary by more than 50°C, even though the reference thermocouple remains at constant temperature. The automatic control system is used to eliminate this 'chimney effect'. The surface in T–T'–χ space is constructed, where T and T' are temperatures at the crystal position and reference thermocouple, respectively. A polynomial form of the equation

$$T' = f(T, \chi)$$

is used in the temperature-control program.

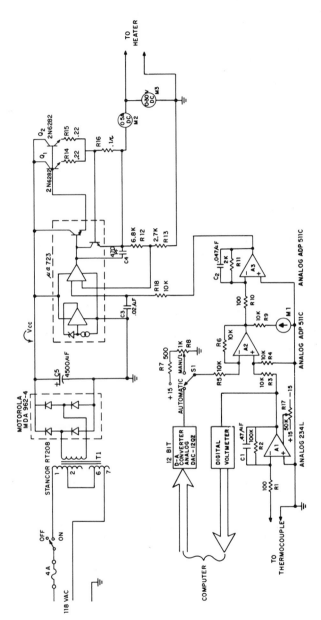

Figure 2-5. Circuit diagram for the programmable furnace controller used in conjunction with an automated diffractometer system. $R1$–$R11$ are metal-film precision resistors. Manufacturers are cited for reference only (from Finger, Hadidiacos and Ohashi, 1973, reproduced by permission of the Carnegie Institution of Washington)

The automatic control system may also be programmed to determine unit-cell dimensions on a crystal, increase or decrease temperature, and repeat the unit-cell dimension measurement as desired. Thermal expansion of a crystal may thus be measured automatically.

II. CRYSTAL MOUNTING

Successful high-temperature crystallography is dependent on a stable crystal mount. Some highly refractory compounds, such as magnesium silicates, may be attached directly to the join bead of a thermocouple. Platinum becomes sticky when heated to incandescence, and a strong bond between crystal and thermo-couple may thus be achieved. Smyth employed this procedure in his studies of Mg_2SiO_4, olivine (Smyth and Hazen, 1973) to 900°C. The obvious advantage of this mounting procedure is its simplicity, although its use is limited to the relatively few compounds that do not alter at high temperature in air. A possible disadvantage is that the thermocouple e.m.f. versus temperature profile may be altered by the presence of a crystal adhering to the join bead, thus adding uncertainty to the temperature calibration. There may also be a time-dependent change in the thermocouple calibration caused by contamination.

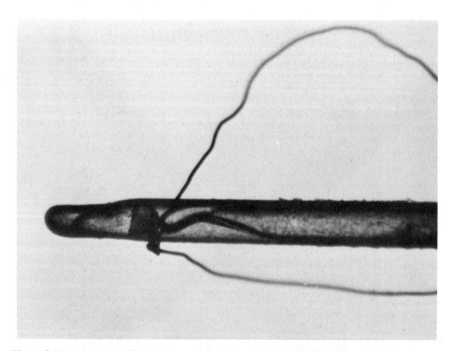

Figure 2-6. A typical silica glass capillary mount with $Pt–Pt_{90}Rh_{10}$ sensing thermo-couple. The crystal of olivine, $(Mg, Fe)SiO_4$, has a maximum dimension of approximately 150 µm. The capillary is evacuated and sealed at both ends to avoid crystal oxidation

By far the most widely used mounting procedure in high-temperature crystallography is sealing of the crystal in a silica glass capillary. The single crystal is placed at the sealed end of a capillary with inner diameter slightly larger than the crystal diameter. The crystal is then wedged against the end of the capillary by insertion of a silica glass needle, with mullite wool if needed. Figure 2-6 is a photograph of a typical high-temperature capillary mount. Many compounds that decompose or oxidize in air at high temperature (for example compounds with iron) must be isolated from the atmosphere, and the capillary must be evacuated and sealed at the end opposite the crystal. This is most easily accomplished with capillaries drawn from, but still attached to, silica tubing. Such tubing is available commercially from Blake Industries, Charles Supper, Inc., or General Rand, Inc. In most experiments to date the simple evacuation of the silica capillary has proved sufficient to preserve crystals from oxidation, based on the

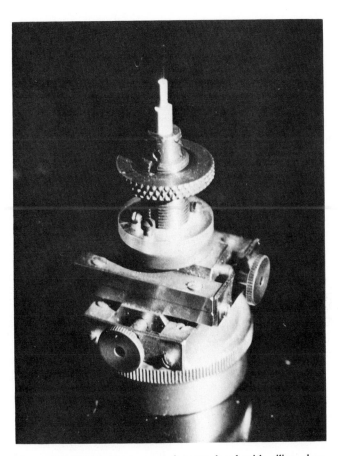

Figure 2-7. Standard x-ray goniometer head with silica glass capillary mount. The height of the assembly is 5 cm

lack of change in unit-cell parameters, colour or composition (Prewitt, 1976). It should be possible to buffer oxygen fugacity using powdered mixtures or different gas environments if more rigorous atmospheric controls are required.

The completed high-temperature mount is usually a thin silica capillary approximately 1 cm in length and 0.4 mm in diameter, with the crystal near one end. This capillary mount may be attached to a slightly modified goniometer head with ceramic cement such as Zircoa Bond (Zirconium Corporation of America) or Sauereisen (Sauereisen Cements Co.). A capillary mount on a goniometer head is illustrated in Figure 2-7.

A common problem with silica-glass capillary mounts is loss of the crystal through fracturing of the glass. This occurrence is frequent after heater failure or sudden cooling of the sample, but may also happen during routine operation at constant high temperature, perhaps owing to glass devitrification, which is enhanced by contact with some crystals. Thus, it is important to determine the composition, size and shape of single crystals *before* data collection to ensure that proper absorption corrections can be made if the crystal is lost.

III. TEMPERATURE CALIBRATION

A vital step in successful high-temperature crystallography is accurate temperature calibration. All calibration procedures currently in use rely on a sensing thermocouple; the geometry and details of measurement, however, differ widely. It should be emphasized that all methods of temperature measurement must first be tested against primary or fixed-point standards. The most reliable fixed points are melting points; several convenient standards with their melting temperatures are listed in Table 2-2. Each individual thermocouple differs owing to the size and interface geometry of the join bead. Each thermocouple should

Table 2-2. Fixed calibration points at high temperature

Material	Melting Point* (°C)
Pb metal	327.4
AgBr	432
AgI	558
NaI	651
KCl	776
NaCl	801
Ag metal	960.8
NaF	988
Au metal	1063
Cu metal	1083

* Melting point data from *Handbook of Chemistry and Physics* R. C. Weast (ed.) Chemical Rubber Co., Cleveland, Ohio.

therefore be calibrated against these fixed-point standards in the appropriate experimental configuration before data are collected.

The most common thermocouple configuration in high-temperature work is a $Pt-Pt_{90}Rh_{10}$ bead adjacent to, but outside, the silica glass capillary (Figure 2-6). This thermocouple may read out directly to a recorder, or it may be used in a feedback system to regulate power to the furnace. In general, the temperature at the thermocouple will not be the same as at the crystal inside the capillary, so careful calibration experiments are required. The thermocouple bead, which is in air, should be as small as possible. Join beads more than 1.5 times the wire diameter may produce low temperature readings owing to thermal conduction away from the bead (Hazen, 1976).

Use of an internal calibrant holds the potential for highly accurate temperature determination, independent of thermocouple measurements. A single crystal of a cubic material with precisely determined thermal expansion coefficient could be included in the glass capillary adjacent to the single crystal under investigation. A single lattice parameter measurement on the standard cubic crystal is sufficient to define temperature; in the case of sodium chloride, if the unit-cell edge is measured to 1 part in 10^4 (the normal precision) then temperature is defined to within 0.7°C. Furthermore, systematic errors in thermocouple calibration are reduced with this internal calibration technique. The importance of internal calibration will be re-emphasized in discussion of simultaneous high-temperature and high-pressure measurements.

IV. FUTURE PROSPECTS

In spite of the continuing introduction of new single-crystal heater designs (e.g. Ishizawa *et al.*, 1978), no substantial improvements in temperature range or stability have been made since the work of Brown, Sueno and Prewitt (1973). The lack of new developments may be in large part due to the vast number of crystallographic problems yet to be studied in the temperature range below 1100°C, although the present capabilities may also reflect the approximate upper limits of radiative heating.

Should higher temperatures be required other heating techniques may prove necessary. In addition to the open-flame procedures mentioned above, laser heating of single crystals may hold promise for generating very high temperatures. As with flame heaters, however, it is difficult to maintain constant uniform temperatures with lasers. A more readily soluble problem is how to improve temperature stability and accuracy of calibration with heaters presently in use. An increase in thermal mass of the heater, coupled with shielding of air currents, should allow reduction of temperature fluctuations, and a corresponding increase in accuracy, to better than ±2°C.

REFERENCES

Brown, G. E., S. Sueno and C. T. Prewitt (1973) A new single-crystal heater for the precession camera and four-circle diffractometer. *Amer. Mineral.* **58**, 698–704.

16

Finger, L. W., C. G. Hadidiacos and Y. Ohashi (1973) A computer-automated, single-crystal, x-ray diffractometer. *Carnegie Inst. Washington Year Book* **72**, 694–699.

Foit, F. F., and D. R. Peacor (1967) A high-temperature furnace for a single-crystal diffractometer. *J. Sci. Instrum.* **44**, 183–185.

Glazer, A. M. (1972) A technique for the automatic recording of phase transitions in single crystals. *J. Appl. Cryst.* **5**, 420–423.

Goldschmidt, H. J. (1964) *Bibliography 1: High-temperature X-ray Diffraction Techniques.* Utrecht, International Union of Crystallography, Commission on Crystallographic Apparatus.

Hazen, R. M. (1976) Effects of temperature and pressure on the crystal structure of forsterite. *Am. Mineral.* **61**, 1280–1293.

Ishizawa, N., T. Miyata, I. Minato and S. Iwai (1978) A high temperature apparatus for the four-circle diffractometer. Tokyo Inst. Technology *Rept. of the Res Lab. of Engineering Materials* **3**, 15–18.

Lissalde, F., S. C. Abrahams and J. L. Bernstein (1978) Microfurnace for single-crystal diffraction measurements. *J. Appl. Cryst.* **11**, 31–34.

Lynch, R. W., and B. Morosin (1971) A hemispherical furnace for high-temperature single-crystal x-ray diffraction studies. *J. Appl. Cryst.* **4**, 352–356.

Prewitt, C. T. (1976) Crystal structures of pyroxenes at high temperature. In *The Physics and Chemistry of Minerals and Rocks*, R. G. Strens (ed.), John Wiley, New York.

Prewitt, C. T., J. J. Papike and M. Ross (1970) Cummingtonite: a reversible nonquenchable transition from $P2_1/m$ to $C2/m$ symmetry. *Earth Planet. Sci. Lett.* **8**, 448–450.

Reeber, R. R. (1975) An automatically recording high-temperature Laue camera. *Z. Kristallogr.* **141**, 465–470.

Smyth, J. R. (1969) Orthopyroxene-high-low clinopyroxene inversions. *Earth Planet. Sci. Lett.* **6**, 406–407.

Smyth, J. R. (1972) A simple heating stage for single-crystal diffraction studies up to 1000°C. *Amer. Mineral.* **57**, 1305–1309.

Smyth, J. R., and R. M. Hazen (1973) The crystal structures of forsterite and hortonolite at several temperatures up to 900°C. *Amer. Mineral.* **58**, 588–593.

Tuinstra, F., and G. M. Fraase Storm (1978) A universal high-temperature device for single-crystal diffraction. *J. Appl. Cryst.* **11**, 257–259.

Viswamitra, M. A., and K. Jayalakshmi (1970) A simple miniature furnace for routine collection of single crystal x-ray data up to 1000°C on the Hilger and Watts linear diffractometer. *J. Phys. E: Sci. Instr.* **3**, 656–657.

Wyckoff, R. W. G. (1925) The crystal structure of the high temperature form of cristobalite (SiO_2). *Am. J. Sci.* series 5, **9**, 448–459.

Chapter 3

High-pressure Crystallography

CONTENTS

I	Development of the Diamond Pressure Cell	18
	A Powder Diffraction	18
	B Single-crystal Diffraction	20
II	Diamond-cell Design	23
	A Types of Single-crystal Pressure Cell	23
	(1) Lever-arm cell	23
	(2) Fourme pneumatic cell	25
	(3) Merrill and Bassett miniature cell	26
	(4) Stuttgart radially symmetric cell	27
	(5) Schiferl transverse-geometry cell	27
	(6) Mao and Bell cryogenically loaded cell	28
	B Construction of a Diamond Cell	29
	(1) Triangular steel supports	29
	(2) Diamond support discs	31
	(3) Diamond anvils	31
	(4) Gaskets	33
III	Operation of the Diamond Cell	33
	A Specimen Selection and Mounting	33
	(1) Choosing the specimen	33
	(2) Diamond mounting and alignment	34
	(3) Mounting and centring the gasket	36
	(4) Mounting the crystal	36
	(5) Pressure fluids and sealing the cell	37
	(6) Raising and cycling the pressure	38
	B Pressure Calibration	38
	(1) The ruby fluorescence technique	38
	(2) Internal pressure standards	39
	C X-ray Photography with the Diamond Cell	40
	(1) Modified goniometer head	40
	(2) X-ray source	40
	(3) High-pressure precession photography	41
	D Diffractometry	46
	(1) Centring the crystal	46
	(2) Unit-cell determination	50
	(3) Diamond-cell absorption	51
	(4) Intensity collection	51

  E   Data Reduction and Refinement Procedures   52
       (1)   Data averaging   52
       (2)   Robust/Resistant refinement   53
       (3)   Quality of results   53
IV   Future Prospects   54

I. DEVELOPMENT OF THE DIAMOND PRESSURE CELL

A. Powder diffraction

Diamond is remarkable, not only because it is the hardest known material, but also because it is highly transparent to many ranges of electromagnetic radiation. These attributes of diamond have led to its present widespread use in the generation of high-pressure environments for physical experimentation. Despite the general recognition of diamond's unique properties, it was not until the late 1940s that Lawson and Tang (1950) at the University of Chicago first employed diamonds in a high-pressure x-ray experiment. The Lawson and Tang device (Figure 3-1), though similar in appearance to some modern diamond cells, was fundamentally different in the geometry of pressure generation. Semi-cylindrical grooves were ground into each half of a cleaved diamond. The diamonds were clamped together, and the resulting cylindrical hole became the high-pressure sample chamber. Two tungsten piano wires inserted into the holes acted as pistons, compressing a polycrystalline sample. Powder x-ray photographs were then taken on the sample at pressures greater than 20 kbar. This earliest diamond

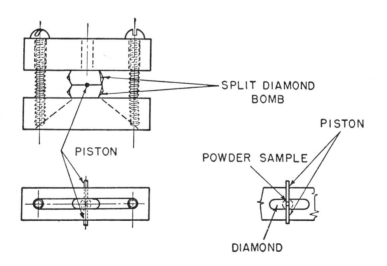

Figure 3-1. Split-diamond pressure cell of Lawson and Tang (1950). (Reproduced by permission of the American Institute of Physics)

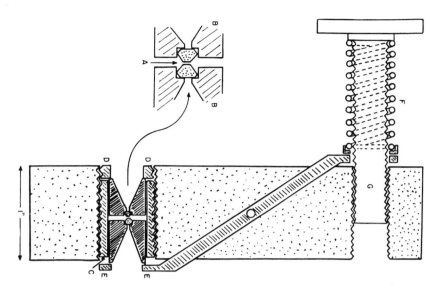

Figure 3-2. Opposed-diamond anvil, lever-arm pressure cell of Weir *et al.* (1959). (A) Diamond anvils, (B) diamond supports, (C) cylinder, (D) stop, (E) lever arm, (F) spring, (G) adjustment screw. (Reproduced by permission of the United States National Bureau of Standards)

cell, like its successors, generated very high pressures by applying a moderate force to a small surface area. Furthermore, because sample volumes are small, and the stored energy available for catastrophic release is low, diamond cells proved to be an extremely safe method of experimenting with high pressure.

John Jamieson (1957), also at the University of Chicago, improved the Lawson and Tang design by drilling a hole directly through a single diamond crystal. The principle of operation was the same as the earlier Chicago device, but the pressure range was increased to greater than 30 kbar with the stronger confining vessel. Using this pressure cell, Jamieson was successful in identifying high-pressure phase transitions in calcium carbonate ($CaCO_3$, calcite) and other materials in his early powder x-ray experiments. Jamieson's success led Charles Weir and Alvin Van Valkenburg of the United States National Bureau of Standards (NBS) to construct a similar cell using an eight carat gem diamond, which had been confiscated by the United States government from smugglers. Workers at General Electric also employed the single drilled diamond cell, with the modification of conical rather than cylindrical holes in the diamonds (Kasper *et al.*, 1960). The drilled diamond cell, though capable of sustaining the highest pressures of any contemporary x-ray device, was severely limited, as was painfully realized when the eight carat NBS diamond shattered at moderate pressure.

The idea of an opposed diamond configuration for generating high pressure was developed independently and nearly simultaneously by workers at the National Bureau of Standards (Figure 3-2, Weir *et al.*, 1959) and the University of Chicago

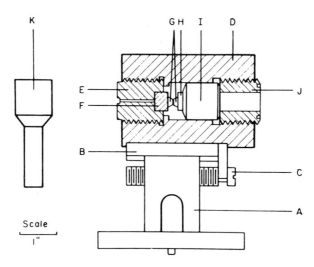

Figure 3-3. Opposed-diamond anvil, high-pressure cell of Jamieson *et al.* (1959). (A–C) translation stage, (D–E, J) threaded cell for pressure application, (F–I) diamond anvils and supports. (Reproduced by permission of the American Institute of Physics)

(Figure 3-3, Jamieson, Lawson and Nachtrieb, 1959). In their earliest experiments the NBS group used the diamond cell as an optical window on the high-pressure environment, viewing samples in transmitted light, whereas workers at Chicago concentrated on powder x-ray diffraction techniques with the x-ray source parallel to the diamond anvil faces. Both groups, however, employed the opposed-diamond geometry, which permits the generation of very high pressures at the diamond interface with comparatively little force, and which is now the basic element of all diamond cells. Both drilled-diamond and opposed-diamond anvil cells were used successfully by many workers for high-pressure powder x-ray diffraction studies, the results of which have been reviewed in part by Munro (1963) and Rooymans (1969).

Powder x-ray diffraction studies have continued to be an important direction in diamond cell research. Diffraction patterns of samples at pressures in excess of 1 Mbar have been obtained (Mao and Bell, 1975; 1978; Mao *et al.*, 1979). Most of the dozens of high-pressure powder studies were undertaken primarily for phase identification and compressibility measurements, and are not considered further in this monograph.

B. Single-crystal diffraction

In its original form the opposed-diamond cell was not suitable for single-crystal x-ray diffraction because crystals would be crushed between the anvil faces. An important development, therefore, was Van Valkenburg's (1964) metal foil gasketing technique in which a thin metal sheet with a hole smaller than the diamond anvil faces was placed between the diamonds to act as a sample chamber (Figure 3-4). The gasket was originally designed to study the crystallization of fluids at high pressure, and the first high-pressure, single-crystal, x-ray experiments were performed on ice VI in a gasketed diamond cell (Block, Weir and Piermarini, 1965; Weir, Block and Piermarini, 1965). There followed several more studies by the National Bureau of Standards group on crystallized liquids, including benzene, bromine, carbon disulphide and carbon tetrachloride (Weir, Piermarini and Block, 1969a; Block, Weir and Piermarini, 1970; Piermarini and Braun, 1973). It was also recognized by the NBS researchers that the gasketing procedure could be used to enclose an initially solid single crystal in a hydrostatic environment. Experiments on high-pressure phases of potassium nitrate (Weir, Piermarini and Block, 1969a) and on elemental caesium and gallium (Weir, Piermarini and Block, 1971) demonstrated the usefulness of this approach.

All of the pioneering NBS single-crystal work was performed with a specially built, modified Buerger precession camera (Weir, Piermarini and Block, 1969b) with a lever-arm pressure cell made entirely of beryllium, save for the gasket and diamond components (Figure 3-5). This diamond cell was too large to be easily adapted to standard x-ray equipment. These early contributions therefore included only high-pressure, unit-cell dimensions and space groups, and did not present complete structure refinements based on integrated intensity data. Merrill and Bassett (1974) applied the opposed-diamond principle in their construction of a miniature diamond cell for single-crystal x-ray diffraction (Figure 3-6). This cell,

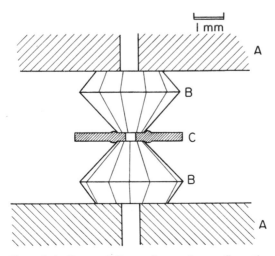

Figure 3-4. Opposed-diamond anvil configuration
with a metal foil gasket
(A) diamond supports, (B) diamond anvils, (C) gasket

22

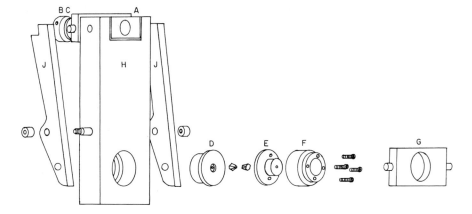

Figure 3-5. Exploded view of the lever-arm, opposed-diamond anvil pressure cell of Weir *et al.* (1969b). The cell is composed of beryllium to minimize x-ray absorption. (Reproduced by permission of the American Institute of Physics)

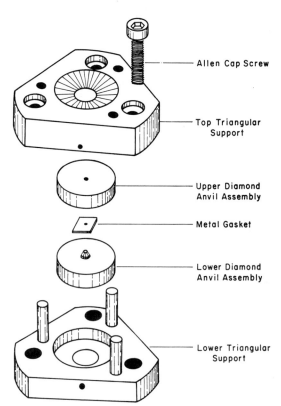

Allen Cap Screw

Top Triangular Support

Upper Diamond Anvil Assembly

Metal Gasket

Lower Diamond Anvil Assembly

Lower Triangular Support

Figure 3.6. Exploded view of the miniature diamond anvil of Merrill and Bassett (1974), as modified by Hazen and Finger (1977). Load is applied by three Allen cap screws

simple in construction and use, is small enough to be adaptable to most single-crystal x-ray cameras and diffractometers. The first complete three-dimensional structure refinements of crystals at high-pressure were performed on $CaCO_3$ (calcite; Merrill and Bassett, 1975) and $BaFeSi_4O_{10}$ (gillespite; Hazen and Burnham, 1974) using the Merrill and Bassett cell, a modification of which is described in detail in the subsequent section.

II. DIAMOND-CELL DESIGN

A. Types of single-crystal pressure cell

All diamond pressure cells now in use for x-ray diffraction employ the opposed-diamond geometry. There are significant differences, however, in the mechanisms for generating load, as well as the orientation of primary and diffracted x-radiation with respect to the diamond anvils.

Several factors must be considered in selecting a mechanism for applying force to the diamonds. The load must be applied gradually and should be continuously variable. Diamond alignment is the single most important factor in obtaining high pressure, and the diamond anvil faces must stay concentric and parallel during loading to prevent severe gasket deformation or diamond breakage. The loading mechanism must not be too bulky, nor should it obstruct the path of primary and diffracted x-rays. It is, of course, desirable to have as simple and inexpensive a design as possible, consistent with the other constraints.

In addition to a number of different diamond support mechanisms, two types of diffraction geometry are available in single-crystal, high-pressure diffraction experiments. Most devices employ the transmission mode in which x-rays pass through one diamond, the single crystal, and then the other diamond. In Jamieson's 1959 diamond cell, on the other hand, a lateral or transverse geometry, in which the x-rays enter and exit through the same diamond, is used (Figure 3-3). Schiferl (1977) also employed a transverse geometry in his novel single-crystal diamond cell (Figure 3-7). Important factors in selecting which geometry to use are the accessibility of reciprocal space and the ease of correcting diffraction data for absorption effects. Several types of single-crystal cells are considered below with respect to loading mechanisms and diffraction geometry.

(1) Lever-arm cell

The NBS lever-arm design (Figure 3-5) has proved to be the most widely adapted in diamond cell research. The first x-ray cell was composed entirely of beryllium to increase the accessible region of reciprocal space. This construction has proved too costly and steel components are now employed in all lever-arm cells. The main advantage of the lever arm is the maintenance of excellent diamond alignment as pressure is raised. Diamond alignment may be optimized in the lever-arm cell by incorporating the diamonds in a piston-cylinder arrangement (Figure 3-8). Lever-arm cells have been operated to above 1.7 Mbar, the highest

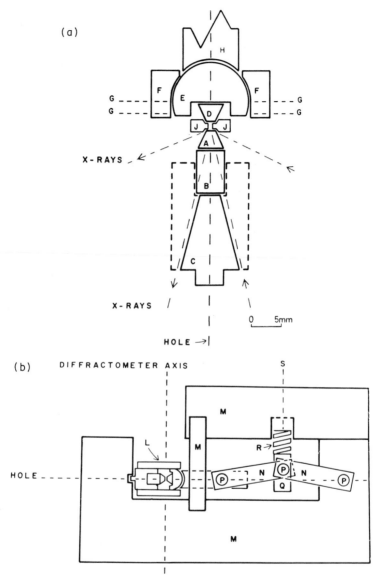

(a)

X-RAYS

X-RAYS

0   5mm

HOLE →|

(b)   DIFFRACTOMETER AXIS

HOLE

Figure 3-7. Diamond-anvil x-ray cell of Schiferl (1977), which uses transverse geometry.
A. Detail of diamond anvil (A and D) and gasket (J) portions of the cell.
B. Assembled cell with loading device
(Reproduced by permission of the American Institute of Physics).

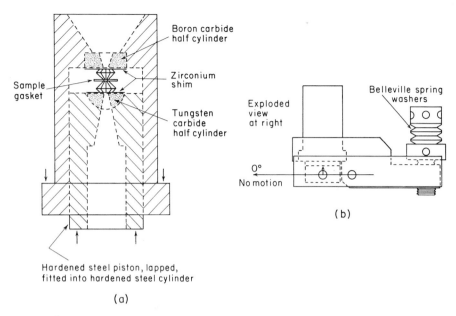

Figure 3-8. Piston cylinder alignment mechanism for the lever-arm, diamond-anvil pressure cell of Mao and Bell (1975).
A. Enlarged view of the piston cylinder and opposed diamonds.
B. Lever-arm support for the piston cylinder mechanism.
(Reproduced by permission of the Carnegie Institution of Washington)

static pressure ever obtained in a laboratory experiment (Mao and Bell, 1978). At present, however, lever-arm cells are not generally used in single-crystal work because of their awkward size and the limited accessibility of reciprocal space in models now available. Furthermore, with the current limit of approximately 100 kbar for most hydrostatic experiments (see below), the enhanced alignment of the lever arm cell is not required. In experiments with liquified or solid forms of a gas such as hydrogen or neon, which are acceptable as hydrostatic pressure media to several hundred kilobars (Mao and Bell, 1979; Hazen et al., 1980a, b), lever-arm devices will undoubtedly be modified for single-crystal experiments. The critical modifications must be miniaturization combined with an increase in the accessible volume of reciprocal space.

(2) Fourme pneumatic cell

One of the first diamond cells designed exclusively for single-crystal x-ray studies was Fourme's (1968) pneumatic cell for use on a Buerger precession camera (Figure 3-9). The principal advantage of the cell is the relatively large volume of reciprocal space available for photographic study. The maximum

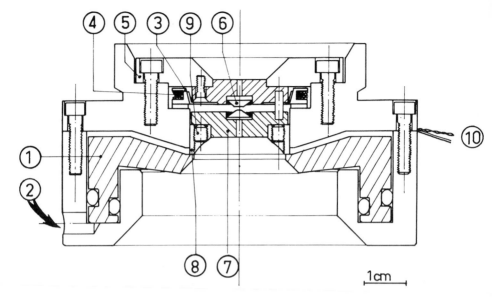

Figure 3-9. Pneumatic pressure cell of Fourme (1968). This device was designed to fit on a Buerger precession camera. (Reproduced by permission of the International Union of Crystallography)

obtainable pressure is approximately 30 kbar, with a pneumatic system applying the load. The Fourme cell is not easily adaptable to apparatus other than the precession camera owing to its size, weight and loading mechanism. A variant of the Fourme pressure cell, designed for growing crystals in a compressed liquid on a precession camera, is commercially available from Enraf Nonius (see Appendix I).

(3)   Merrill and Bassett miniature cell

The miniature diamond cell of Merrill and Bassett (1974; Figure 3-6) is one of the most compact, least expensive, and simplest to operate, and is consequently the most widely used of all single-crystal pressure devices. The light weight and small size of the miniature cell has made it easily transferable between most standard x-ray diffraction equipment. The cell may be mounted on a slightly modified goniometer head (see below) and thus may be treated as a standard crystal mount in many experiments. The radially symmetric design combines ease of use in transmission mode with accessibility to a relatively large volume of reciprocal space. The load is applied by three independently tightened screws at the corners of the triangular cell.

The compact size and simple operation of the Merrill and Bassett cell are gained at the expense of precise diamond alignment. In the original design the problem of poor diamond alignment led to a pressure limit of about 45 kbar and occasional diamond breakage. Modified alignment pins and operating procedures (see below), however, have increased the safe operating range of the miniature cell to above 70 kbar. For most high-pressure x-ray experiments the Merrill and Bassett design is the most effective pressure cell available. Details of the construction and operation of this device are presented in section II-B.

(4) Stuttgart radially symmetric cell

Keller and Holzapfel (1977) designed a single-crystal x-ray cell that is similar to the Merrill and Bassett cell with respect to accessibility to reciprocal space, but which has improved diamond alignment and a more uniform loading mechanism, allowing pressure generation to about 100 kbar (Figure 3-10). The Stuttgart cell is significantly more massive than the miniature design, as well as being far more complex in construction. The added pressure range available with this design must be weighed against the increased expense and operating difficulties.

(5) Schiferl transverse-geometry cell

David Schiferl (1977) continued the tradition of early workers at the University of Chicago by designing a diamond cell with transverse x-ray geometry (Figure 3-7). This design, unique among single-crystal cells, has the great advantage of accessibility to nearly half of the Ewald sphere. High-angle diffraction positions can be measured, for example, resulting in potentially increased precision in lattice constant measurements. The original design was limited to about 50 kbar, but modifications of the cell have allowed experimentation at pressures near the 100 kbar hydrostatic limit.

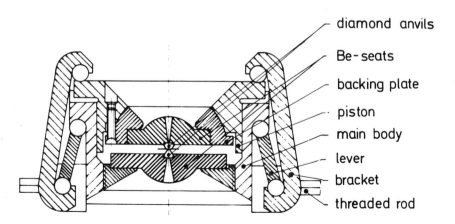

Figure 3-10. Single-crystal diamond-anvil pressure cell of Keller and Holzapfel (1977). (Reproduced by permission of the American Institute of Physics)

Although the Schiferl cell provides excellent access for unit-cell determinations, the transverse geometry is not well suited for integrated intensity measurements. This limitation is a consequence of the very complex absorption paths of x-rays passing through the side of a diamond, gasket, thin crystal, and back through the diamond, as compared with the simple radial symmetry of absorption in transmission mode experiments.

The Schiferl transverse geometry cell is an important complementary high-pressure, single-crystal, x-ray device. For example, in many materials such as graphite or layer silicates only flat platy crystals are available. Thus, certain crystallographic directions are always near perpendicular, whereas other directions are near parallel to the diamond anvil faces. A combination of transverse and transmitted x-ray geometries ensures maximum sampling of reciprocal space.

(6) Mao and Bell cryogenically loaded cell

Mao and Bell (1980) constructed a diamond cell for the study of single crystals of materials that are normally gases at ambient conditions. The cell is similar in principle to the Merrill and Bassett (1974) design, though extensive modifications are required. The cell is circular, with four load-bearing screws, rather than the triangular three-screw configuration of the earlier pressure device (Figure 3-11). In

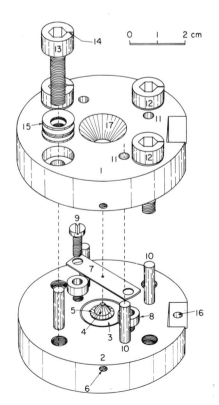

Figure 3-11. Modified diamond-window, high-pressure cell for study of single crystals of solidified gases. (1, 2) end plates, (3) beryllium support discs, (4) diamond anvils, (5) epoxy cement, (6) set screws for adjusting part 3, (7) gasket, (8) spacers, (9) screw, (10) positioners, (11) hole for positioners, (12, 13) screws for raising sample pressure, (14) groove, (15) Belleville springs, (16) mounting notch. (From Mao and Bell, 1980, reproduced by permission of the Carnegie Institution of Washington)

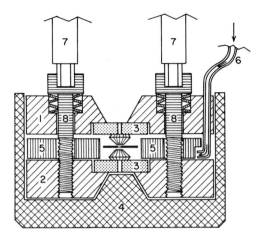

Figure 3-12. Mao and Bell cryogenically loaded cell ready for experiment (see also Figure 3-11). (1, 2) end plates, (3) beryllium support discs, (4) cup, (5) spacers, (6) gas tube, (7) extensions for rotating screws (8). (Reproduced by permission of the Carnegie Institution of Washington)

the Mao and Bell device opposite screws are oppositely threaded, and screws are tightened in pairs, in order to maintain diamond alignment. Loading of a liquified sample at cryogenic conditions is accomplished in a dewar (Figure 3-12), and the cell is closed and pressure sealed remotely with four 2 m long Allen wrenches.

The first experiments with this pressure cell, on single-crystal argon, neon and methane (Hazen *et al.*, 1980a, b; Finger *et al.*, 1981), revealed that these solids are extremely soft even above 100 kbar and might make excellent hydrostatic pressure media. Neon has recently been employed in x-ray studies of single-crystal FeO to 220 kbar (Hazen *et al.*, 1981), which is the highest pressure thus far obtained in a single-crystal experiment. The cryogenically loaded cell has thus opened a new range of conditions for single-crystal x-ray diffraction studies.

B. Construction of a diamond cell

The Merrill and Bassett type cell, with minor modifications by Hazen and Finger (1977), has proved to be the most versatile and least expensive of all available single-crystal, x-ray diamond cells. Details of the construction and use of this cell are presented below. Completely assembled cells may be purchased from High-Pressure Diamond Optics, Inc., and from Blake Industries (see Appendix I).

(1) Triangular steel supports

Two triangular steel supports are machined from hardened steel 1.5 in rod stock (Figure 3-13). An important modification from the original design of Merrill

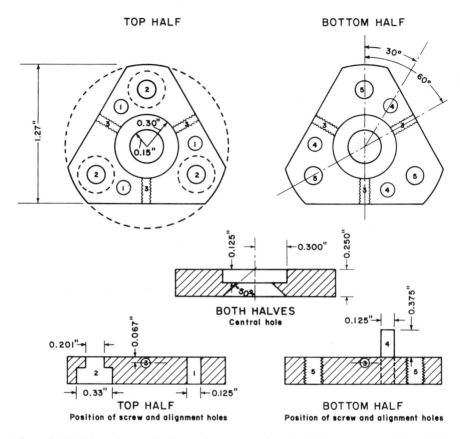

Figure 3-13. Triangular steel diamond supports for the high-pressure and the high-temperature/high-pressure diamond-anvil cells (modified from Merrill and Bassett, 1974). Both halves have three-fold symmetry. (1) 0.318 cm hole for guide pin, (2) well for 10-32 Allen cap screw, (3) tap for 3-48 set screw, (4) guide pin 0.318 cm diameter, (5) tap for 10-32 screw. (Reproduced by permission of the American Institute of Physics from Hazen and Finger, 1981)

and Bassett (1974) is the use of three 0.25 in diameter guide pins, rather than the two 1/16 in pins of the original cell. These pins serve to maintain diamond alignment, thus increasing the available pressure and reducing the chances of diamond fracture. The 50 degree angle of access represents a compromise between maximized strength of support and maximum availability of reciprocal space. For low pressure experiments (below 20 kbar), wider access angles, perhaps up to 70 degrees, could be used, if appropriate modifications are also made to the gasket and diamond anvil components. If routine operation to above 70 kbar is desired, an access angle of less than 50 degrees will provide better diamond support and increased lifetime of beryllium diamond backing discs.

The two triangular supports are held together, and load is applied, with three standard 0.5 in 10–32 Allen cap screws.

(2) Diamond support discs

Diamonds must be supported by a strong material that is transparent to x-radiation. In the original Merrill and Bassett cell support discs were machined from nuclear-grade beryllium, and this practice has been followed in most subsequent applications. The original disc was 0.5 in diameter, with a 0.05 in diameter central hole for optical access, and a 'well' holding the diamond. These discs are satisfactory for operation to 40 kbar, but the discs fail at higher pressures owing to the relatively low tensile strength of beryllium metal. For experiments to 60 kbar nuclear-grade discs with 1/32 in holes and no indentation perform satisfactorily (Figure 3-14). Owing to the extreme toxicity of Be metal it is strongly recommended that parts be machined by a specialist in the metal. Speedring Manufacturers and Beryllium Corporation of America are two American suppliers (see Appendix I).

For pressures greater than 60 kbar alternative diamond supports may be required. In some experiments optical access to the sample may not be essential (see Chapter 4) and in such cases solid Be discs with no hole should be adequate to 100 kbar. If, however, the sample must be observed visually at high pressure, a modified pre-stressed beryllium seat has been constructed by David Schiferl for use with the Merrill and Bassett cell (Figure 3-15). Pre-stressed Be components possess nearly three times the tensile strength of ordinary metal. The hemispherical design thus combines the added support of a thicker backing with a stronger material. Although this type of diamond support is not recommended for routine operation of the Merrill and Bassett cell, because of its increased absorption of x-rays, the hemispherical supports are valuable for the highest pressure range of the Merrill and Bassett cell.

(3) Diamond anvils

Relatively little experimentation on alternative diamond anvil designs has been attempted for the Merrill and Bassett cell; virtually all cells use brilliant-cut

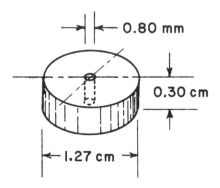

Figure 3-14. Beryllium diamond support discs for use in the modified Merrill and Bassett (1974) pressure cell

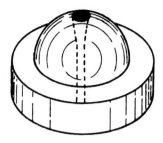

Figure 3-15. Modified beryllium diamond supports, designed by D. Schiferl, for use
with the Merrill and Bassett
(1974) pressure cell to pressures
above 60 kbar.

diamonds of 10 to 15 points (i.e. 0.10 to 0.15 carat) with anvil faces from 0.6 to
1.0 mm in diameter (Figure 3-16). The smaller faces are used in higher-pressure
experiments. No special crystallographic orientation of diamonds is required, nor
is the type of diamond important in x-ray work.

The size and shape of diamond anvils might be modified for special cases. If
low pressures, below 20 kbar, are of interest then anvil faces may be polished to
larger than 1.0 mm. This procedure yields correspondingly thinner diamond anvils
and should allow larger sample volume with less x-ray absorption by the
diamonds. Larger anvil faces, combined with larger gasket hole diameter (see
below) will be essential if access angles greater than 50 degrees are attempted in
the triangular supports. Best results are obtained if the two anvils are closely
matched in diameter. Diamond dealers who supply matched anvil pairs can
usually provide custom shaped diamonds for special applications. Dubble Dee
Corporation and Lazare Kaplan and Sons are firms that supply diamond anvils for
high-pressure studies (see Appendix I).

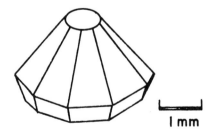

l mm

Figure 3-16. Diamond anvil for high-
pressure studies. Diamonds are usually
0.10–0.15 carat gems with anvil faces
from 0.6 to 1.0 mm in diameter

(4)   Gaskets

The gasket is a piece of thin sheet metal with a small hole drilled in it. The gasket is thus the simplest component of the diamond-anvil single-crystal cell, yet in many ways it is the most critical. This circumstance is due to the fact that most diamond cell experiments terminate at high pressure in extreme deformation (i.e. failure) of the gasketed sample chamber. (Gasket failure is of course less costly than failure of diamond or beryllium components of the cell!) Selection of appropriate gasket material and dimensions is thus an important step in assembling the diamond cell.

For most applications the gasket is made of a hard metal foil approximately 250 μm thick. Commonly used foils are Inconel 750X (International Nickel Company) foil 0.01 in thick, which is a moderately hard nickel alloy, and T301 stainless steel foil 0.01 in thick, which is a hard steel. The outer dimensions of the gasket are not critical and will depend in part on the cell geometry, though a 1 cm square is convenient for many applications.

The size of the gasket hole depends on the size of the crystal under study and the maximum desired pressure. In general, the hole must be at least three times the sample crystal diameter in order to prevent shielding of x-rays. An optimum diameter for the sample is 100 μm, with a gasket hole of 350 μm. The gasket hole must be smooth and even, with no burr. One method for producing consistently accurate holes is to machine drill gasket blanks with #80 drill holes, and then use a hand held pin vise with #79 or #78 drills to achieve the desired hole diameter, combined with emery paper to remove any burr around the gasket hole.

Several workers have experimented with preindented gaskets, which are impressed by the diamonds prior to drilling the gasket holes. Although this procedure may increase the available pressure range in some experiments, the difficulty of centring the hole after indenting, coupled with the decrease in available sample volume after indenting, limits the effectiveness of the technique. A great deal of experimentation with alternative gasket materials and geometries has yet to be done, making further developments in gasketing one of the most fruitful grounds for advances in high-pressure, single-crystal technology.

III. OPERATION OF THE DIAMOND CELL

This section is intended as a step-by-step guide to the aligning and operation of the Merrill and Bassett miniature diamond cell. A list of tools and supplied is given in Table 3-1.

A. Specimen selection and mounting

(1)   Choosing the specimen

Single crystals for high-pressure x-ray diffraction must be small enough to remain uncrushed between the diamond anvils, and unshielded from x-rays by the gasket, yet large enough to provide sufficient diffracting volume for intensity measurements. In practice, a flat plate $100 \times 100 \times 40$ μm is ideal for most

Table 3-1. Supplies required for mounting crystals in the Merrill and Bassett (1974) single-crystal, diamond-anvil pressure cell, as modified by Hazen and Finger (1977)

Tools	Materials
1. Binocular microscope with variable magnification and top and bottom illumination	1. Epoxy or Duco type cement
	2. Jewellers' wax
2. Pin vise with #78 and #79 drills (approximately 300 μm diameter)	3. Vaseline petroleum jelly
	4. Acetone
3. Sharp needle for crystal manipulation	5. Kimwipes or other soft tissue
4. Allen wrenches	6. Emery cloth
5. Diamond cell, including triangular supports, beryllium discs, two diamonds and metal gasket	7. Methanol: ethanol in 4:1 ratio
	8. Ruby chips ground to 5–10 μm
	9. Sample single crystal

materials with linear absorption coefficients less than $100 \, \text{cm}^{-1}$. Dense, heavy-atom materials with higher absorption may necessitate thinner crystals (e.g. Schiferl, 1977); some light-atom crystals, on the other hand, require thicker samples (e.g. Hazen and Finger, 1979). In the latter case maximum obtainable pressure is limited by the crystal thickness.

Many crystals have natural cleavage or parting and flat plates are easily obtained. In other materials, or in specimens for which a specific orientation is desired, flat plates may be produced by embedding a grain in epoxy resin and grinding to the desired thickness.

(2) Diamond mounting and alignment

Precise and reproducible diamond alignment is the most important factor in obtaining high pressure with the diamond cell. Misaligned diamonds cause premature gasket failure, and may result in diamond breakage. With nominal care, however, satisfactory diamond alignment is easily achieved and maintained. This operation, as well as many subsequent aspects of crystal mounting, is most easily performed with the aid of a variable-power binocular microscope with both upper and lower light sources.

The diamonds must first be cemented to the beryllium discs. Two cements commonly used are a duco-type, acetone-soluble glue, and epoxy resin. Epoxy is preferable because it is more permanent and less likely to loosen during diamond cleaning and crystal mounting. The objective of diamond mounting is to centre the gems precisely over the support disc holes, without getting glue into the holes. One successful procedure is to anchor each diamond over a hole with a small drop of duco cement at the side of the diamond. When dry, a girdle of epoxy cement is sculpted around the diamond. This produces a strong bond, without obscuring the optical access through the cell.

Two beryllium discs with diamonds attached are now inserted into the triangular steel supports. The two diamond and disc assemblies are clamped into

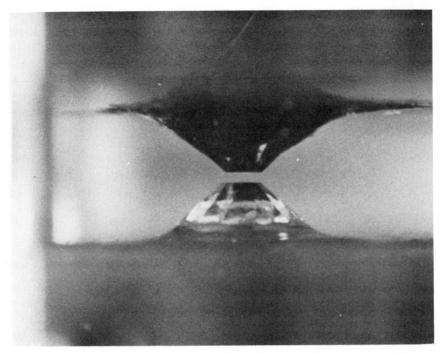

Figure 3-17. Photomicrograph of opposed diamonds mounted in a Merrill and Bassett-type cell. Precise alignment of the anvil faces is critical to attain high pressure and prevent diamond breakage. Anvil faces are 0.7 mm in diameter

the triangular support using three set screws (Figure 3-13). These three screws are used to adjust the lateral position of the upper diamond anvil to superimpose precisely on the lower anvil. This operation is most easily performed looking at the side of the anvils with a binocular microscope (Figure 3-17). Note that once aligned, care should be taken always to reassemble the cell in the same relative orientation of the two triangular halves.

Aligned diamonds will generally remain aligned unless one gem breaks loose. However, visual confirmation of alignment is strongly recommended prior to each crystal mounting.

An alternative method of gasket preparation and mounting is to centre the hole as noted above, but then clamp down on the gasket until the hole closes down to only 100 μm diameter. The gasket is then removed and the hole redrilled to the original ≃350 μm diameter. The gasket, which has clear octagonal impressions of both upper and lower diamonds, is recentred over the lower diamond. Though more time-consuming, this redrilling procedure has the advantage of allowing higher pressures with less gasket deformation or shielding by the pre-thinned gasket.

(3) Mounting and centring the gasket

The gasket must be carefully centred and secured against the lower diamond anvil. This is easily accomplished by first placing three 3 mm diameter balls of jeweller's wax on the beryllium seat around the diamond. A small speck of dust is placed at the precise centre of the otherwise clean diamond anvil face. The gasket is pressed onto the three wax balls with the gasket hole exactly centred over the speck of dust. In this way the gasket is centred.

In order to prevent leakage of pressure fluid when the cell is first closed the gasket must be slightly impressed onto the bottom diamond after centring. Close the two halves together (without fluid) and tighten the three load screws slightly —just enough to indent the top surface of the gasket. When reopened, the gasket should reveal a light impression of the top anvil exactly centred over the gasket hole (Figure 3-18). The diamond cell is now ready for crystal mounting.

(4) Mounting the crystal

Crystals are secured to the lower diamond face using the alcohol-insoluble fraction of petroleum jelly. A small drop of this 'glue' is placed on the lower diamond face, and an additional smear is applied to the side of the gasket hole with a sharp needle. The flat single crystal is centred in the gasket hole, using the needle as a manipulator. Ruby calibration crystals (see below) about 10 µm in diameter are attached to the sides of the gasket hole, which is now ready to be filled with fluid and sealed. In mounting the crystal, care should be taken not to embed the single crystal completely in petroleum jelly—a minimum of adherent should be used to keep the crystal in a hydrostatic environment. In addition, the ruby chips should be kept to the sides of the single crystal rather than on top of the crystal, to prevent premature crushing of the sample.

It is occasionally of interest to compare the pressure variations of two or three crystals of similar composition. In this circumstance it may be advantageous to prepare a multiple-crystal mount, in which the crystals may be compared at identical pressures. By removing pressure as an unknown, small relative changes

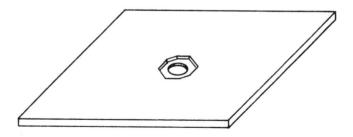

Figure 3-18. Metal foil gasket after indentation by diamonds. The gasket hole and the diamond faces should be concentric. The gasket is 1 cm square

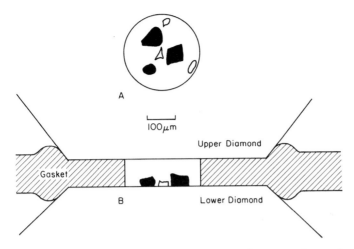

Figure 3-19. Compound crystal mount. Two or three crystals may be included in the same mount in order to compare pressure variations of related materials (A) top view, (B) side view

can be detected. A compound mount is prepared in much the same way as a conventional mount, with the two or three crystals arranged about the centre of the gasket hole (Figure 3-19).

(5) Pressure fluids and sealing the cell

Methanol plus ethanol in 4 : 1 ratio has become the most widely used fluid pressure medium in diamond cells since the discovery by Piermarini, Block and Barnett (1973) that the mixture is hydrostatic to above 100 kbar. Recent work by Piermarini (personal communication) has demonstrated that dried alcohols are less effective than normal water-bearing alcohols. The optimum ratio of methanol:ethanol:water has been found by Piermarini to be 16:3:1. For all experiments above 30 kbar this methanol–ethanol mixture is now the recommended pressure fluid. Two disadvantages of methanol–ethanol may be avoided in experiments below 30 kbar, in which range glycerin is the preferred hydrostatic medium. Methanol–ethanol is more difficult to use than glycerin because of the tendency for a bubble to form and get trapped in the sample chamber, as well as the rapid evaporation of the former fluid. The formation of bubbles in the alchohol may be reduced by first 'washing' the sample chamber, ruby chips and crystal in acetone. Alcohols appear to wet the surfaces of gasket and crystals more thoroughly when they are first cleaned in this manner. After fluid is added to the gasket hole area any trapped bubbles are removed with a sharp needle or fine hypodermic syringe, and the diamond cell is closed and tightened. Small residual bubbles, which are almost unavoidable in the case of methanol–ethanol, are quickly dissolved in the fluid at low pressure (less than 1 kbar) and do not significantly affect the experiment.

Mao and Bell (1979) have recently experimented with liquified hydrogen and helium, and Hazen *et al.* (1980a, b) and Finger *et al.* (1981) have studied crystallized argon, neon and methane, which may serve as hydrostatic pressure media to several hundred kilobars in the cyrogenically loaded cell. These solidified 'gases' are likely to facilitate the next major advance in high-pressure single-crystal studies.

(6) Raising and cycling the pressure

The single-crystal x-ray cell is now sealed and ready to be pressurized. Pressure is raised by simply turning each of the three load screws in sequence by 5 to 10 degree increments. Owing to the very different compressive properties of the gasket and pressure transmitting fluid, the sample chamber will generally compress in a non-uniform way. For example, using a 350 µm hole in 250 µm thick Inconel 750X foil, with methanol–ethanol as the hydrostatic pressure medium, the hole compresses concentrically to about 250 µm in diameter and 150 µm thick at 30–35 kbar. With additional increments of the load screws pressure does *not* increase; rather the hole diameter expands and the gasket becomes thinner as the Inconel gasket can no longer support the internal fluid pressure. Continued tightening of the cell would result in gasket failure with no appreciable increase in pressure.

Higher pressures may be obtained by 'cycling' the pressure. The cell is tightened only to the point at which the gasket hole diameter is reopened to the original 350 µm diameter (although at this point it will be only about 100 µm thick). Pressure is released to the point at which a small amount of fluid is seen to escape rapidly. On retightening, the gasket hole once again is observed to shrink in diameter, and the pressure may be raised to above 50 kbar before another cycle is required. Because a thin gasket reduces x-ray shielding, it is standard procedure to cycle the pressure at least once before performing x-ray experiments.

It is in principle possible, given fortuitous selection of fluid medium, gasket material and gasket hole size, to maintain a constant hole diameter while gradually raising pressure. Such a configuration might prolong gasket life and allow higher pressures. Sufficient data on gasketing variables do not exist, and, for a time at least, gasketing will remain more an art than a science.

B. Pressure calibration

(1) The ruby fluorescence technique

The development at the NBS of a high-pressure calibration procedure based on the wavelength shift of the R_1 fluorescence line of ruby (Forman *et al.*, 1972) transformed the diamond cell into a quantitative instrument. The shift is nearly linear to pressures above 1 Mbar (Bell and Mao, 1979) and may be measured to a

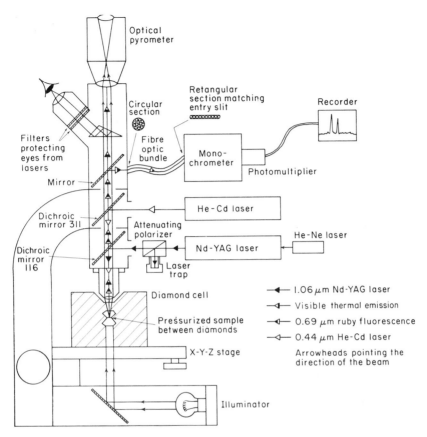

Figure 3-20. Schematic diagram of a ruby fluorescence pressure calibration system (Mao and Bell, 1975). This design also incorporates a YAG laser for heating capability. (Reproduced by permission of the Carnegie Institution of Washington)

precision equivalent to 0.5 kbar or better (Barnett, Block and Piermarini, 1973; Bell and Mao, 1975; King and Prewitt, 1980). Essential elements of a calibration system are a high-intensity light source, such as a laser, to excite the fluorescent radiation, and a sensitive spectrometer to measure the wavelength (Figure 3-20). Several laboratories employ a laser Raman spectrometer for rapid and routine pressure calibration, whereas others have built dedicated calibration systems. Details of a calibration system particularly suited to single-crystal x-ray studies have been presented by King and Prewitt (1980).

(2) Internal pressure standards

The ruby pressure calibration is ideal for its speed and ease of application. Approximate pressures (± 2 kbar) can be determined on the system that we use in about 1 minute, and more careful measurements take no more than 15 minutes. The ruby method, however, is inherently limited to a precision of a few tenths of a kilobar. An alternative calibration procedure of far greater potential precision is the use of an internal x-ray standard single crystal.

Unit-cell dimensions can be measured to a precision of 1 part in 10^4 or better with present diffractometry techniques. For a cubic alkali halide such as NaCl, with bulk modulus of $\simeq 250$ kbar, knowledge of the cell edge to 0.01 per cent implies knowledge of pressure to ± 0.05 kbar or better. Internal calibration, though time-consuming, thus holds the promise of greatly enhanced precision of pressure determination.

Even more precise pressure calibration should be possible with a crystallized rare gas, such as argon or neon. Hazen *et al.* (1980a, b) and Finger *et al.* (1981) have demonstrated that these gases crystallize readily in the gasketed cell to single crystals, which recrystallize rather than deform as the gasket changes shape. Argon or neon may thus be used as a simultaneous pressure medium and calibration. The cubic form of Ar and Ne above crystallization at 12 kbar and 55 kbar, respectively, have among the greatest compressibilities of any known solids. Unit-cell length measurements of argon and neon to 1 part in 10^4 could provide a pressure calibration sensitive to ± 5 bars at 50 kbar!

C. X-ray photography with the diamond cell

*** WARNING ***

EXTREME CAUTION should be exercised while using the diamond cell in an x-ray beam owing to the high level of scattered radiation, and the possibility of strongly directed diffraction off the diamond single crystals.

*** !!! ***

(1) Modified goniometer head

Standard x-ray goniometer heads are easily modified to hold the miniature Merrill and Bassett cell. A pressure cell cradle (Figure 3-21) has been designed to attach directly onto heads by Huber or Supper. The triangular pressure cell is fastened to the cradle with two small screws in the side of the lower triangular support. The cell is easily removed for ruby calibration and can be reattached to the goniometer head without loss of crystal orientation.

Alternatively, a specially designed goniometer head with z-axis translation and more massive cell support has been designed at the Geophysical Laboratory (Hazen and Finger, 1977) as illustrated in Figure 3-22.

(2) X-ray source

Molybdenum $K\alpha$ radiation is widely used for high-pressure x-ray diffraction because of its low absorption by diamond and beryllium metal. A silver target

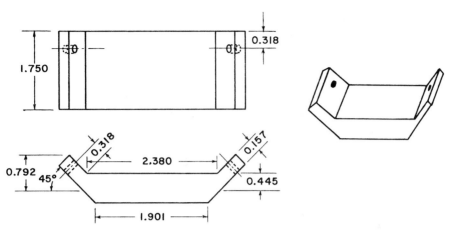

Figure 3-21. Cradle support system of the Merrill and Bassett (1974) diamond pressure cell. The cradle may be adapted to most types of commercially available goniometer heads (see also Figure 3-22)

Figure 3-22. Goniometer head, cradle, and pressure cell for single-crystal x-ray diffraction

could have higher transmission factors; however, the intrinsic brightness of silver targets is lower. In addition the short wavelength increases problems of multiple diffraction. It should be noted that the accepted wavelength of Mo Kα radiation has recently been modified from 0.70926 A to 0.70930 A (*International Tables for X-ray Crystallography*, Vol. **4**). This change is significant if unit-cell dimensions are determined to better than 1 part in 10^4, and the value of x-ray wavelength assumed should thus be reported in all crystallographic studies.

The use of monochromated x-radiation greatly enhances x-ray photographs with the diamond cell. However, monochromated radiation is not essential in most applications.

(3) High-pressure precession photography

The Merrill and Bassett cell attached to a modified goniometer head will fit on most standard single crystal x-ray cameras. Because of the geometry of diffraction through the cell, however, the precession camera is best suited for crystal orientation and photography of reciprocal space (Figure 3-23).

The diamond cell is easily centred in the plane perpendicular to the x-ray beam on a precession camera by shining a light through the diamond cell and aligning this light with the x-ray collimator. Centring of the cell parallel to the x-ray beam is accomplished by making the separation from beryllium to the end of the

Figure 3-23. Single-crystal pressure cell on a precession camera

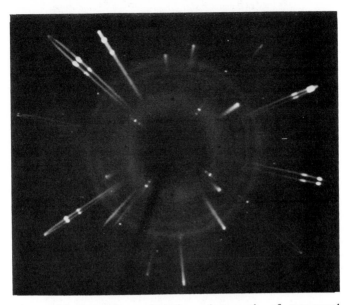

Figure 3-24. Precession orientation photograph of tetragonal MnF_2 (rutile form) at 10 kbar and 21°C. Visible on the photograph are powder rings from beryllium, strong streaks from the two diamond anvils, and sharp diffraction maxima from the $100 \times 100 \times 40$ μm crystal of MnF_2 in the (001) orientation. The photograph was taken with Polaroid Type 57 film, Mo radiation, 30 minutes at 45 kV and 15 ma

collimator equidistant in both standard and 180 degrees from standard orientation. (In all diamond cell work at the Geophysical Laboratory 'standard' orientation is defined such that the screws point in the same direction as the primary x-ray beam). On most precession cameras a C-shaped beam stop is required owing to the size of the cell.

Polaroid orientation photographs are routinely taken of all single crystal high-pressure mounts. Three major diffraction effects are seen on these photographs: diamond streaks, beryllium powder rings, and sharp single-crystal spots (Figure 3-24). The single-crystal effects, though weaker than diamond and beryllium diffraction, are usually discernible by their sharpness and lack of correspondence with diamond or beryllium effects. Contrast between single-crystal diffraction and background scattered radiation may be greatly enhanced through the use of a layer-line screen to mask non-zero-level diffraction.

In most experiments a single orientation photograph is sufficient to achieve orientation on a diffractometer. Occasionally, however, precession photographs prove informative regarding high-pressure transformations and twinning (Hazen, 1976; Figure 3-25). Another useful technique on the precession camera is cone-axis photography (Figure 3-26), which may reveal changes in lattice spacings perpendicular to the plane of the diamond anvil faces. In general, however,

44

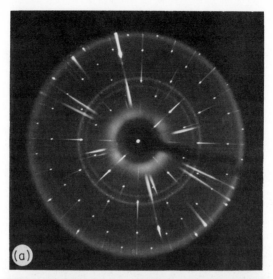

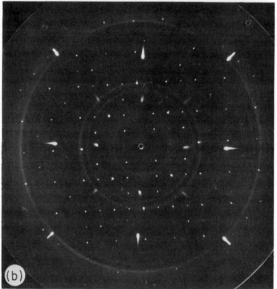

Figure 3-25. Precession photographs of gillespite,
$BaFeSi_4O_{10}$.
A. Tetragonal gillespite I at 12 kbar. Visible on the
photograph are beryllium powder rings, intense diamond
streaks and sharp spots from gillespite in the (001)
tetragonal orientation. The photograph was taken with
Ilford G film, Zr-filtered Mo radiation, 24 hours at 40 kV
and 16 ma.
B. Orthorhombic gillespite II at 20 kbar in the (001)
orientation. The photograph was taken with Ilford G film,
monochromated Mo K radiation, exposed for 48 hours at
35 kV and 15 ma

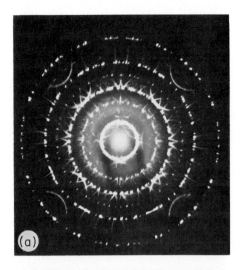

Figure 3-26. Cone axis photographs of the c axis of gillespite, $BaFeSi_4O_{10}$.

A. Gillespite I at room conditions. The 0- to 6-levels are present, corresponding to a 16 A repeat distance.

B. Gillespite II at 27 kbar. The 0-, 1- and 2-levels are present, corresponding to an 8 A spacing of the high-pressure form.

Both photographs were taken with Ilford G film, Zr-filtered Mo radiation, 12 hour exposures at 40 kV and 16 ma

diffractometery has proved the most effective method of characterizing the lattice geometry of crystals at high pressure.

D. Diffractometry

(1) Centring the crystal

A major advantage of the use of single-crystal diffractometry to characterize lattice geometry at high pressure is that the position of the diffraction may be measured in three dimensions in reciprocal space, in contrast to the single dimension of powder techniques. Assignment of Miller indices to diffraction intensity is less ambiguous and interaxial angles may be more directly observed.

The definition of diffractometer angles used here is the same as that given by Busing and Levy (1967), in which the incident and diffracted beams define a horizontal plane. The counter also lies in this plane and may be rotated about a vertical axis to make an angle of 2θ with the incident beam. Mounted on the instrument is a goniostat (see Figure 2-4) consisting of a full circle with two motions: ϕ, which rotates the crystal about an axis lying in the circle, and χ, which rotates the ϕ-axis around the circle. The goniostat is rotated about a vertical axis to an angle θ when the detector is rotated 2θ. This setting is the bisecting position. The goniostat may be rotated independently of the detector to an angle ω, which is the deviation from bisecting position. When χ is zero, the crystal carrier mechanism is at the bottom of the circle. In this position, the 2θ, ω and ϕ-axes are colinear and left-handed rotations of these axes are chosen as the positive direction. A right-handed motion for χ is chosen as positive.

The intensity-centring algorithm used in our laboratory is similar to the technique described by Busing (1970). The diffracting condition is established by rocking the crystal via a series of step scans on ω, with the maximum intensity chosen as the peak position.

After an ω position is selected, the counting time required for 3 per cent relative error (1111 counts) is computed and this time is used for all further counting intervals. The vertical centring of the beam in the detector is accomplished by changing χ to locate the angles at which 25 per cent of the maximum intensity is above or below the counter aperture. A half-interval search technique is used to speed this procedure. After χ is centred, the best value for 2θ is found by a similar half-interval search method moving the detector (2θ) but holding the crystal fixed in position. For diffractometers with $\theta-2\theta$ mechanical coupling, compensating drives on ω are required to perform this operation; ϕ remains fixed throughout the centring sequence. For crystals of maximum dimension less than 0.01 cm, the angles located by the centring procedure are reproducible to approximately 0.01 degrees.

The apparent position of the diffracted radiation is affected by many types of systematic errors. Included among these are uncertainties in diffractometer zero positions, the errors in crystal centring and misalignment of the instrument. Hamilton (1974) and King and Finger (1979) devised a procedure whereby the

Table 3-2. Indices, angles and crystal displacements for reflections used in crystal centring

Setting No.	Indices			Angles				Displacement		
1	h	k	l	2θ	ω	χ	ϕ	Δx	Δy	Δz
2	$\bar{h}$	$\bar{k}$	$\bar{l}$	2θ	ω	$-\chi$	$\pi + \phi$	$-\Delta x$	$-\Delta y$	Δz
3	h	k	l	-2θ	$-\omega$	$\pi + \chi$	ϕ	Δx	$-\Delta y$	$-\Delta z$
4	$\bar{h}$	$\bar{k}$	$\bar{l}$	-2θ	$-\omega$	$\pi - \chi$	$\pi - \phi$	$-\Delta x$	Δy	$-\Delta z$
5	h	k	l	2θ	$-\omega$	$\pi - \chi$	$\pi + \phi$	$-\Delta x$	Δy	$-\Delta z$
6	$\bar{h}$	$\bar{k}$	$\bar{l}$	2θ	$-\omega$	$\pi + \chi$	ϕ	Δx	$-\Delta y$	$-\Delta z$
7	h	k	l	-2θ	ω	$-\chi$	$\pi + \phi$	$-\Delta x$	$-\Delta y$	Δz
8	$\bar{h}$	$\bar{k}$	$\bar{l}$	-2θ	ω	χ	ϕ	Δx	Δy	Δz

measured angles for a single reflection, or Friedel pair of reflections, in eight different orientations are used to determine the values for the errors described above. In addition, an estimate of the angles corrected for these effects is also obtained. When the corrected values are used to refine the lattice parameters, much higher precision results. In high-pressure experiments, where crystal centring by optics is difficult or impossible, these techniques are invaluable.

Table 3-2 lists the settings used to obtain eight equivalent reflections on a diffractometer with limited ω motion. The crystal displacement is also shown for a coordinate system with x along the x-ray beam, z along the rotation axis away from the goniometer head when χ is zero, and y is chosen to make a right-handed system with x and z. Table 3-3 lists the contributions of each of the various terms to the observed angles $T = 2\theta_{obs}$, $D = \omega_{obs} + \theta_{obs}$, and $A = \chi_{obs}$. Note that the factors in Table 3-2 and 3-3 are established for a diffractometer with the parity conditions for a Picker unit, i.e. left-handed 2θ, ω, and ϕ, and right-handed χ. In Table 3-3 the terms with subscript t are the 'true' angles, corrected for the errors; the terms with subscripts of 0 are the errors in the zero settings for the instrument; $\Delta\chi_{xt}$ is the error in χ arising from an error in counter or tube height; and C_x, C_y, S_x and S_y relate errors in x and y to changes in T and D.

The unknown quantities in the expressions of Table 3-3 may be found from appropriate linear combinations of the expressions. After solutions for these terms have been computed, the crystal offsets Δx_θ, Δy_θ, and Δz_θ may be found. These offsets are in the θ-coordinate system and must be transformed into the ϕ-system that is associated with the goniometer head. It is also necessary to know R_c and R_s, the crystal-to-counter aperture and crystal-to-source distances. The former can be obtained from the manufacturer's specifications or from a direct measurement; however, R_s depends on the beam divergence and is strongly dependent upon the characteristics of any monochrometer present. It is suggested that R_s be measured by displacing an optically centred crystal along y and determining C_y and S_y. The source-to-crystal distance is given by $R_s = C_y R_c / S_y$. The uncertainty associated with this calculation is large and we recommend that several determinations be made. Furthermore, R_s should be remeasured whenever there are

Table 3-3. Contributions to observed angles for settings used in crystal centring

Setting No.	T	D	A
1	$2\theta_t + 2\theta_0 - C_x - C_y + S_x - S_y$	$\omega_t + \omega_0 + \theta_t + S_x - S_y$	$\chi_t + \chi_0 + \Delta\chi_{xt} - \Delta\chi_c$
2	$2\theta_t + 2\theta_0 + C_x + C_y - S_x + S_y$	$\omega_t + \omega_0 + \theta_t - S_x + S_y$	$-\chi_t + \chi_0 + \Delta\chi_{xt} - \Delta\chi_c$
3	$-2\theta_t + 2\theta_0 - C_x - C_y + S_x - S_y$	$-\omega_t + \omega_0 - \theta_t - S_x - S_y$	$\chi_t + \chi_0 + \Delta\chi_{xt} + \Delta\chi_c + \pi$
4	$-2\theta_t + 2\theta_0 + C_x + C_y - S_x + S_y$	$-\omega_t + \omega_0 - \theta_t + S_x + S_y$	$-\chi_t + \chi_0 + \Delta\chi_{xt} + \Delta\chi_c - \pi$
5	$2\theta_t + 2\theta_0 + C_x - C_y - S_x + S_y$	$-\omega_t + \omega_0 + \theta_t - S_x - S_y$	$-\chi_t + \chi_0 - \Delta\chi_{xt} - \Delta\chi_c - \pi$
6	$2\theta_t + 2\theta_0 - C_x + C_y + S_x + S_y$	$-\omega_t + \omega_0 + \theta_t + S_x + S_y$	$\chi_t + \chi_0 - \Delta\chi_{xt} - \Delta\chi_c + \pi$
7	$-2\theta_t + 2\theta_0 + C_x - C_y - S_x + S_y$	$\omega_t + \omega_0 - \theta_t + S_x - S_y$	$-\chi_t + \chi_0 - \Delta\chi_{xt} + \Delta\chi_c$
8	$-2\theta_t + 2\theta_0 - C_x + C_y + S_x + S_y$	$\omega_t + \omega_0 - \theta_t - S_x + S_y$	$\chi_t + \chi_0 - \Delta\chi_{xt} + \Delta\chi_c$

Table 3-4. Observed angles and calculated parameters

Setting	T	D	A	ϕ
1	46.902	22.862	−75.735	−25.695
2	46.986	22.869	75.555	−205.695
3	−46.868	−22.793	104.215	−25.695
4	−46.989	−22.8535	−104.420	−205.695
5	46.900	24.055	−104.305	−205.695
6	46.904	24.068	104.410	−25.695
7	−46.893	−24.0315	75.645	−205.695
8	−46.923	−24.0535	−75.535	−25.695

$$
\begin{aligned}
2\theta_t &= (T_1 + T_2 - T_3 - T_4 + T_5 + T_6 - T_7 - T_8)/8 & = 46.921 \\
2\theta_0 &= (T_1 + T_2 + T_3 + T_4 + T_5 + T_6 + T_7 + T_8)/8 & = 0.002 \\
\omega_t &= (D_1 + D_2 - D_3 - D_4 - D_5 - D_6 + D_7 + D_8)/8 & = -0.604 \\
\omega_0 &= (D_1 + D_2 + D_3 + D_4 + D_5 + D_6 + D_7 + D_8)/8 & = 0.015 \\
\chi_t &= (A_1 - A_2 + A_3 - A_4 - A_5 + A_6 - A_7 + A_8 - 4\pi)/8 & = -75.640 \\
\chi_0 &= (A_1 + A_2 + A_3 + A_4 + A_5 + A_6 + A_7 + A_8)/8 & = -0.021 \\
\Delta\chi_c &= (-A_1 - A_2 + A_3 + A_4 - A_5 - A_6 + A_7 + A_8)/8 & = -0.0025 \\
\Delta x_{xl} &= (A_1 + A_2 + A_3 + A_4 - A_5 - A_6 - A_7 - A_8)/8 & = -0.0750 \\
S_x &= (D_1 - D_2 - D_3 + D_4 - D_5 + D_6 + D_7 - D_8)/8 & = -0.0041 \\
S_y &= (-D_1 + D_2 - D_3 + D_4 - D_5 + D_6 - D_7 + D_8)/8 & = -0.0078
\end{aligned}
$$

$$
\Delta h = -\frac{2R_c \sin\theta \Delta\chi_c}{\cos\omega} = \frac{(2)(215)(.39812)(-0.0025)}{(.99994)(57.2958)} = 0.007 \text{ mm}
$$

$$
\Delta x_\theta = \frac{S_x R_s}{\sin\theta} = \frac{(-0.0041)(215)}{(.39812)(57.2958)} = -0.039 \text{ mm}
$$

$$
\Delta y_\theta = \frac{-S_y R_s}{\cos\theta} = \frac{-(-0.0078)(215)}{(0.91733)(57.2958)} = 0.032 \text{ mm}
$$

$$
\Delta z_\theta = \frac{2\sin\theta \Delta\chi_{xl}}{\cos\omega\left(\dfrac{1}{R_s}+\dfrac{1}{R_c}\right)} = \frac{(2)(.39812)(-.0750)}{(.99994)\left(\dfrac{1}{215}+\dfrac{1}{215}\right)(57.2958)} = -0.112 \text{ mm}
$$

$$
R = \begin{pmatrix}
(\cos\omega\cos\phi - \sin\omega\sin\phi\cos\chi) & (\cos\omega\sin\phi + \sin\omega\cos\chi\cos\phi) & (-\sin\omega\sin\chi) \\
(-\sin\omega\cos\phi - \cos\omega\sin\phi\cos\chi) & (-\sin\omega\sin\phi + \cos\omega\cos\chi\cos\phi) & (-\cos\omega\sin\chi) \\
(-\sin\chi\sin\phi) & (\sin\chi\cos\phi) & (\cos\chi)
\end{pmatrix}
$$

$$
= \begin{pmatrix}
0.89993 & -0.43591 & -0.01021 \\
0.11703 & 0.21891 & 0.96870 \\
-0.42003 & -0.87296 & 0.24801
\end{pmatrix}
$$

$$
X_\phi = R'X_\theta
$$
$$
\Delta x_\phi = 0.89993\Delta x_\theta + 0.11703\Delta y_\theta - 0.42003\Delta z_\theta = 0.016 \text{ mm}
$$
$$
\Delta y_\phi = -0.43591\Delta x_\theta + 0.21891\Delta y_\theta - 0.87296\Delta z_\theta = 0.121 \text{ mm}
$$
$$
\Delta z_\phi = -0.01021\Delta x_\theta + 0.96870\Delta y_\theta + 0.24801\Delta z_\theta = 0.004 \text{ mm}
$$

changes that could affect the beam divergence. If the user is only interested in the 'true' angles and not in crystal offsets, R_s need not be known.

Table 3-4 lists the observed angles and the resulting corrected angles and diffractometer errors for a reflection on a Picker instrument with $R_c = R_s = 215$ mm. The equations for calculation of the various terms are also given in the table.

(2) Unit-cell determination

There are many ways that the reflection-centring information can be used to refine the lattice parameters of a crystal. One way is to use measured 2θ values as input to an ordinary lattice constant refinement program. Such a procedure is not recommended because it makes no use of the three-dimensional nature of the data (i.e., angular relationships between the diffraction planes are ignored.) The Oak Ridge diffractometer system (Busing, 1970), on the other hand, used observed diffractometer angles to refine lattice constants and orientation (a total of nine parameters) for a triclinic crystal. Constraints associated with higher symmetries can be easily invoked; however, the calculations are nonlinear and multiple cycles are required. On a slow machine the calculation may require several hours for completion.

An alternate computational scheme has been proposed by Shoemaker and Bassi (1970) and Tichý (1970), who used the generalized diffractometer equation

$$\mathbf{h}_\phi = \mathbf{UBh} \tag{3-1}$$

in which $\mathbf{U}$ is an orthogonal matrix relating the diffractometer axes or ϕ-system to the crystal cartesian axes, $\mathbf{B}$ is the transformation from crystal to crystal cartesian system, $\mathbf{h}$ is a vector with Miller indices as components and $\mathbf{h}_\phi$ is the reciprocal lattice vector in the ϕ-system. For an indexed reflection, $\mathbf{h}$ and $\mathbf{h}_\phi$ are known. If matrices $\mathbf{H}_\phi$ and $\mathbf{H}$ are constructed with columns composed of $\mathbf{h}_\phi$ and $\mathbf{h}$, the general observation equations become:

$$\mathbf{H}_\phi = \mathbf{UBH} \tag{3-2}$$

If $\mathbf{H}_\phi$ and $\mathbf{H}$ have three or more columns, equation (3-2) may be postmultiplied by the transpose of $\mathbf{H}$ to yield:

$$\mathbf{H}_\phi \mathbf{H}' = \mathbf{UBHH}' \tag{3-3}$$

The product $\mathbf{HH}'$ is a 3×3 matrix that will be nonsingular if the reciprocal lattice vectors in $\mathbf{H}$ are noncoplanar. Postmultiplying equation 3-3 by the inverse of $\mathbf{HH}'$, we obtain:

$$\mathbf{UB} = \mathbf{H}_\phi \mathbf{H}(\mathbf{HH}')^{-1} \tag{3-4}$$

The product $\mathbf{UB}$ is the orientation matrix required to obtain the diffractometer angles for any reflection using equation (3-1). The lattice constants are found by premultiplying $\mathbf{UB}$ by its transpose. The result is the reciprocal metric tensor with the ijth element equal to the scalar product $\mathbf{b}_i \mathbf{b}_j$, where the $\mathbf{b}$'s are reciprocal lattice vectors.

This technique has the advantage that only a single 3 × 3 matrix must be inverted, and, in addition, only one iteration need be performed. The method is, therefore, well suited for slow computers. A disadvantage is that it is not easy to include constraints on the lattice parameters and each cell is refined as though it were triclinic. The approach of the resulting lattice to the values required by symmetry is a check on the precision of the experiment.

Shoemaker and Bassi (1970) also presented equations for application of lattice constraints. Although this technique requires inversion of matrices of rank greater than nine, it appears to be very useful. Ralph and Finger (1982) have written a computer program implementing these equations.

(3) Diamond-cell absorption

Intensities measured in the diamond cell must be corrected for absorption by diamond and beryllium components of the cell. For the cell developed at the National Bureau of Standards, Santoro *et al.* (1968) developed a technique for correction of absorption in complex geometries. Denner *et al.* (1978) presented a scheme for absorption correction for their diamond cell, which also has a complex absorption function. Both of these methods utilize time-consuming calculations.

The Merrill and Bassett (1974) diamond cell is a radially symmetric design. Thus the absorption correction is reduced to a one-dimensional problem. Finger and King (1978) measured the relative attenuation of the direct beam and presented a plot of the variation of absorption as a function of the angle, ψ, between the beam and the axis of radial symmetry. If the diamonds and beryllium discs were flat plates, a simple exponential relationship between transmission and the path length would be correct; however, the hole in the disc complicates the results. At least two different correction methods have been used. The first is to use a least-squares technique to determine the coefficients of a polynomial in (sec $\psi - 1$), which describes the log of the attenuation in terms of ψ. A second method is to fit a series of line segments to the attenuation versus ψ curve, which was measured experimentally. Both methods match the profile within a few per cent, are equally valid, and can be programmed for on-line correction of the diamond-cell absorption. If the computer program that controls the experiment cannot be modified for this correction, the additional correction step must be added to a data reduction program that follows.

(4) Intensity collection

The integrated intensities from a crystal mounted in the diamond cell can be measured with data collection sequences identical to those used at room pressure; however, the amount of data accessible is increased if modified sequences are used. In most diamond cells, only 30–40 per cent of reciprocal space is accessible and peak-to-background is reduced. Thus, it is important to measure as many reflections as possible. If a reflection is to be measured, both the incident and diffracted beams must lie within a particular angle of the radially symmetrical axis of the cell (42° in the Merrill and Bassett cell). Otherwise the beam path is through highly absorbing steel of the cell body or gasket. As shown by Finger and King (1978), the sum of these angles for the primary and diffracted beam is a minimum if the radial axis of the cell lies in the equatorial plane of this instrument. This condition is achieved by keeping $\phi = 0$; hence, the so-called fixed-ϕ mode of data collection is used.

Additional changes in data collection procedures arise from the effects of the extra material in the beam. When diffraction from one of the diamonds overlaps one of the crystal reflections, the datum must be discarded; however, the interference from the beryllium may be avoided. These discs are cryptocrystalline and give rise to a series of diffuse powder rings. When a conventional θ–2θ scan is undertaken, these rings may affect the integrated intensity by adding to the peak or by augmenting either the upper or lower background; thus, the appropriate regions of 2θ must be skipped. Alternatively, if an ω-scan is performed, the detector is not moved during the intensity measurement. The background will be increased by the powder ring from the beryllium; however, the contribution will be constant for each reflection and it is not necessary to skip any regions of reciprocal space.

E.   Data reduction and refinement procedures

(1)   Data averaging

In spite of the efforts to obtain high-quality intensity data, the results from high-pressure experiments are less precise than the results from normal experiments. The sources of the errors are many and include accidental and undetected scattering from diamonds and shadowing from irregular features of the gasket. Some of these effects may be detected by careful observations of peak shapes as seen on the strip chart recorder. Other effects can be minimized by measuring all accessible reciprocal space and averaging the symmetrically equivalent reflections. In the triclinic case, only Friedel pairs will have been collected and it is not possible to determine which reflection is aberrant if the two do not agree.

In higher symmetry classes, there are some types of reflection for which more than two observations may be made. For such crystals it is possible to detect data which are internally inconsistent. The first step in this process is to identify symmetry-related reflections and calculate a mean structure factor and the standard deviation of the mean. If the latter value is less than a critical value, usually chosen as 5 per cent of the mean, no further processing is done. For reflections with a standard deviation greater than this value, the point that has the largest discrepancy from the mean is tentatively rejected and a new mean and standard deviation are computed. The process is repeated until the reflection has a significantly small standard deviation or until only two reflections remain. In the latter case, the rejection algorithm has failed and the data must be manually treated. The above scheme is designed to function on distributions that consist of clusters of data with a few outlier points.

(2) Robust/resistant refinement

The averaging procedure described above will facilitate the detection and rejection of many of the observations with incorrect structure factors; however, it is possible for incorrect values to be included in the data set submitted to least-squares analysis. If these defective values are included in the observational equations at full weight, then they may cause the refinement to diverge. On the other hand, one must be careful not to bias the results by exclusion of results that disagree with the current model, which is not necessarily correct.

A means of resolving this dilemma has been suggested by Prince *et al.* (1977) and Prince (1978) and is called the robust/resistant technique. A method of adjustment of variable parameters is said to be robust if it yields good estimates of the parameters even though the distribution of errors in the observation is not Gaussian and/or if the data points are incorrectly weighted. The term resistant is applied if the fitting technique is insensitive to the presence of a small number of data that are inconsistent with the rest. For diamond-cell intensities both robust and resistant qualities are needed.

Robustness is added by a dynamic modification of the weighting factors. For each reflection, the term $\Delta F/\sigma$, which is the square root of the contribution to the weighted residual, is calculated. This factor is divided by nine times the median delta over sigma, to yield a dimensionless parameter x that modifies the weighting according to the relationship:

$$w' = w(1 - 2x^2 + x^4) \tag{3-5}$$

where w' is the revised weight and w is the ordinary weighting factor. Resistance to badly measured data is added by setting $w' = 0$ for $|x| > 1$.

Equation (3-5) is a function that will change smoothly from w for observations that exactly match the model, to 0 for widely discrepant values. As refinement proceeds, the median delta becomes smaller and the revised weights are adjusted correspondingly. The refinement is accomplished without subjective identification of bad observations because of the automated procedure. In addition, as the parameters of the model are improved by adding an extinction coefficient, for example, many data thought originally to be discrepant will be returned to the unrejected state. It is our opinion that the satisfactory refinement of diamond-cell intensity data is very difficult without the use of these robust/resistant techniques and our refinement program RFINE4 (Finger and Prince, 1975) has been modified to incorporate these calculations.

(3) Quality of results

If the precautions described above are followed, it is usually possible to refine a data set to R-factors that are roughly equivalent to those of a data set measured at room pressure. The results for the high-pressure set, however, will not be as precise because the number of observations is reduced by a factor of two to four, thereby increasing the estimated standard deviations by factors as large as two.

The data are also collected within a restricted region of reciprocal space and not distributed randomly. The correlations between parameters may thus be increased compared to those of room-condition refinements. Furthermore, the estimated standard deviations on calculated parameters, such as bond distances and angles, may be increased by factors of 3 or 4 relative to ordinary refinement. The loss of precision is unfortunate because the purpose of many high-pressure experiments is to determine the differences between bond distances at two different pressures.

IV. FUTURE PROSPECTS

High-pressure crystallography is in its infancy, and many instrumental improvements will be made in the near future. Diamond-cell design could be improved by optimizing the size and shape of diamond anvils and beryllium support discs for maximum strength and minimum x-ray absorption. Steel supports that allow greater access to reciprocal space could be devised. Gaskets constructed from different metals, or of different size, may facilitate the attainment of high pressure and use of larger sample volumes. Every component of the diamond cell should be considered in the development of improved apparatus.

New types of hydrostatic pressure media, including the rare gases and hydrogen, will provide low-stress environments at significantly higher pressures than the 100 kbar limit now imposed by liquid media. In conjunction with these 'gas media', techniques for routine loading of samples at cryogenic conditions will be required.

Pressure calibration is another experimental procedure in need of improvement. Present techniques yield pressure measurements accurate to only ± 0.5 kbar at 50 kbar. The use of an internal single-crystal x-ray standard might provide calibration better then 0.01 kbar at 50 kbar. In summary, virtually all aspects of the preceding discussion are subject to modification and improvement.

REFERENCES

Barnett, J. D., S. Block and G. J. Piermarini (1973) An optical fluorescence system for quantitative pressure measurement in the diamond-anvil cell. *Rev. Scien. Instrum.* **44**, 1–9.

Bell, P. M., and H. K. Mao (1975) Laser optical system for heating experiments and pressure calibration of the diamond-window high-pressure cell. *Carnegie Inst. Washington Year Book* **74**, 399–402.

Bell, P. M., and H. K. Mao (1979) Absolute pressure measurements and their comparison with the ruby fluorescence (R_1) pressure scale to 1.5 Mbar. *Carnegie Inst. Washington Year Book* **78**, 665–669.

Block, S., C. E. Weir and G. J. Piermarini (1965) High-pressure single-crystal studies of ice VI. *Science* **148**, 947–948.

Block, S., C. E. Weir, and G. J. Piermarini (1970) Polymorphism in benzene, naphthalene, and anthracene at high pressure. *Science* **169**, 586–587.

Busing, W. R. (1970) Least-squares refinement of lattice and orientation parameters for use in automatic diffractometery. In *Crystallographic Computing*, F. R. Ahmed (ed.), Munksgaard, Copenhagen, 319–330.

Busing, W. R., and H. A. Levy (1967) Angle calculations for 3- and 4-circle x-ray and neutron diffractometers. *Acta Cryst.* **22**, 457–464.

Denner, W., H. Schulz and H. D'Amour (1978) A new measuring procedure for data collection with a high-pressure cell on an x-ray four-circle diffractometer. *J. Appl. Cryst.* **11**, 260–264.

Finger, L. W., R. M. Hazen, G. Zou, H. K. Mao and P. M. Bell (1981) Structure and compression of crystalline neon and argon at high pressure and room temperature. *Appl. Phys. Lett.*, **39**, 892–894.

Finger, L. W., and H. E. King (1978) A revised method of operation of the single-crystal diamond cell and refinement of the structure of NaCl at 32 kbar. *Amer. Mineral.* **63**, 337–342.

Finger, L. W., and E. Prince (1975) A System of Fortran IV Computer Programs for Crystal Structure Computations. *U.S. Natl. Bur. Stand. Tech.* Note **854**, Washington, DC, 129 pp.

Forman, R. A., G. J. Piermarini, J. D. Barnett and S. Block (1972) Pressure measurement made by the utilization of ruby sharp-line luminescence. *Science* **176**, 284.

Fourme, R. (1968) Appareillage pour études radiocristallographiques sous pression et à température variable. *J. Appl. Cryst.* **1**, 23–30.

Hamilton, W. C. (1974) Angle settings for four-circle diffractometers. In *International Tables for X-ray Crystallography*, **4**, Kynoch Press, Birmingham, 273–284.

Hazen, R. M. (1976) Sanidine: predicted and observed monoclinic-to-triclinic reversible transformations at high pressure. *Science* **194**, 105–107.

Hazen, R. M., and C. W. Burnham (1974) The crystal structure of gillespite I and II: a structure determination at high pressure. *Amer. Mineral.* **59**, 1166–1176.

Hazen, R. M., and L. W. Finger (1977) Modifications in high-pressure, single-crystal diamond-cell techniques. *Carnegie Inst. Washington Year Book* **76**, 655–656.

Hazen, R. M., and L. W. Finger (1979) Linear compressibilities of $NaNO_2$ and $NaNO_3$. *J. Appl. Phys.* **50**, 6826–6828.

Hazen, R. M., and L. W. Finger (1981) High-temperature diamond-anvil pressure cell for single-crystal studies. *Rev. Scien. Instrum.* **52**, 75–79.

Hazen, R. M., H. K. Mao, L. W. Finger and P. M. Bell (1980a) Crystal structures and compression of Ar, Ne and CH_4 at 20 °C to 90 kbar. *Carnegie Inst. Washington Year Book* **79**, 348–351.

Hazen, R. M., H. K. Mao, L. W. Finger and P. M. Bell (1980b) Structure and compression of crystalline methane at high pressure and room temperature. *Appl. Phys. Lett.* **37**, 288–289.

Hazen, R. M., H. K. Mao, L. W. Finger and P. M. Bell (1981) Irreversible unit-cell volume changes of wustite single crystals quenched from high pressure. *Carnegie Inst. Washington Year Book* **80**, 274–277.

Jamieson, J. C. (1957) Introductory studies of high-pressure polymorphism to 24,000 bars by x-ray diffraction with some comments on calcite II. *J. Geology* **65**, 334–343.

Jamieson, J. C., A. W. Lawson and N. D. Nachtrieb (1959) New device for obtaining x-ray diffraction patterns from substances exposed to high pressure. *Rev. Scien. Instrum.* **30**, 1016–1019.

Kasper, J. S., J. E. Hilliard, J. W. Cahn and V. A. Phillips (1960) *Research and Development on the Effects of High Pressure and Temperature on Various Elements and Binary Alloys.* Washington: Wright Air Development Center, Technical Report AD244–767.

Keller, R., and W. B. Holzapfel (1977) Diamond anvil device for x-ray diffraction on single crystals under pressures up to 100 kilobar. *Rev. Scien. Instrum.* **48**, 517–523.

King, H. E., and L. W. Finger (1979) Diffracted beam crystal centering and its application to high-pressure crystallography. *J. Appl. Cryst.* **12**, 374–378.

King, H. E., and C. T. Prewitt (1980) Improved pressure calibration system using the ruby R_1 fluorescence. *Rev. Scien. Instrum.* **51**, 1037–1039.

Lawson, A. W., and T. Y. Tang (1950) A diamond bomb for obtaining powder at high pressures. *Rev. Scien. Instrum.* **21**, 815.

56

Mao, H. K., and P. M. Bell (1975) Design of a diamond-window high-pressure cell for hydrostatic pressures in the range 1 bar to 0.5 Mbar. *Carnegie Inst. Washington Year Book* **74**, 402–405.

Mao, H. K., and P. M. Bell (1978) High-pressure physics: sustained static generation of 1.36 to 1.72 Megabars. *Science* **200**, 1145–1147.

Mao, H. K., and P. M. Bell (1979) Observations of hydrogen at room temperature (25 °C) and high pressure (to 500 kilobars). *Science* **203**, 1004–1006.

Mao, H. K., and P. M. Bell (1980) Design and operation of a diamond-window, high-pressure cell for the study of single crystal samples loaded cryogenically. *Carnegie Inst. Washington Year Book* **79**, 409–411.

Mao, H. K., P. M. Bell, K. J. Dunn, R. M. Chrenko and R. C. DeVries (1979) Absolute pressure measurements and analysis of diamonds subjected to maximum static pressures of 1.3–1.7 Mbar. *Rev. Scien. Instrum.* **50**, 1002–1009.

Merrill, L., and W. A. Bassett (1974) Miniature diamond anvil pressure cell for single crystal x-ray diffraction studies. *Rev. Scien. Instrum.* **45**, 290–294.

Merrill, L., and W. A. Bassett (1975) The crystal structure of $CaCO_3$ II, a high-pressure metastable phase of calcium carbonate. *Acta Cryst.* **B31**, 343–349.

Munro, D. C. (1963) Structural determinations by x-rays of systems at high pressure. In *High-Pressure Physics and Chemistry*, R. S. Bradley (ed.), Academic Press, 311–323.

Piermarini, G. J., and A. B. Braun (1973) Crystal and molecular structure of CCl_4 III: a high pressure polymorph at 10 kbar. *J. Chem. Phys.* **58**, 1974–1982.

Piermarini, G. J., S. Block and J. D. Barnett (1973) Hydrostatic limits in liquids and solids to 100 kbar. *J. Appl. Phys.* **44**, 5377–5382.

Prince, E. (1978) A test of robust/resistant refinement on synthetic data sets. *Amer. Crystallogr. Assoc. Programs and Abstracts* **6**, series 2, 37.

Prince, E., W. L. Nicholson and J. A. Buchanan (1977) A reanalysis of the data from the single crystal intensity project. *Amer. Crystallogr. Assoc. Programs and Abstracts* **5**, series 2, 67.

Ralph, R. L., and L. W. Finger (1982) A computer program for refinement of crystal orientation matrix and lattice constants from diffractometer data with lattice symmetry constraints. *J. Appl. Cryst.* **15**, In press.

Rooymans, C. J. M. (1969) The behavior of some groups of chalcogenides under very-high-pressure. In *Advances in high-pressure research* **2**, R. S. Bradley (ed.), 1–96.

Santoro, A., C. E. Wier, S. Block and G. J. Piermarini (1968) Absorption corrections in complex cases. Application to single crystal diffraction studies at high pressure. *J. Appl. Cryst.* **1**, 101–107.

Schiferl, D. (1977) 50-kilobar gasketed diamond anvil cell for single-crystal x-ray diffractometer use with the crystal structure of Sb up to 26 kilobars as a test problem. *Rev. Scien. Instrum.* **48**, 24–30.

Shoemaker, D. P., and G. Bassi (1970) On refinement of the crystal orientation matrix and lattice constants with diffractometer data. *Acta Cryst.* **A26**, 97–101.

Tichý, K. (1970) A least-squares method for the determination of the orientation matrix in single-crystal diffractometry. *Acta Cryst.* **A26**, 295–296.

Van Valkenburg, A. (1964) Diamond high pressure windows. *Diamond Research* 17–20.

Weir, C. E., S. Block and G. J. Piermarini (1965) Single crystal x-ray diffraction at high pressures. *J. Res. Natl. Bur. Stand. (U.S.)* **C69**, 275–281.

Weir, C. E., E. R. Lippincott, A. Van Valkenburg and E. N. Bunting (1959) Infrared studies in the 1- to 15-micron region to 30,000 atmospheres. *J. Res. Natl. Bur. Stand. (U.S.)* **63A**, 55–62.

Weir, C. E., G. J. Piermarini and S. Block (1969a) Crystallography of some high-pressure forms of C_6H_6, CS_2, Br_2, CCl_4, and KNO_3. *J. Chem. Phys.* **50**, 2089–2093.

Weir, C. E., G. J. Piermarini and S. Block (1969b) Instrumentation for single crystal x-ray diffraction at high pressures. *Rev. Scien. Instrum.* **40**, 1133–1136.

Weir, C. E., G. J. Piermarini and S. Block (1971) On the crystal structures of Cs II and Ga II. *J. Chem. Phys.* **54**, 2768–2770.

Chapter 4

High-temperature, High-pressure Crystallography

CONTENTS

	A	Early Work	58
	B	Heating Single Crystals in the Diamond Cell	59
II		Design of a High-temperature Diamond Cell	60
	A	Diamond Supports	60
	B	Gaskets	60
	C	The Miniature Heater	60
	D	Insulation	61
III		Operation of the High-temperature Diamond Cell	64
	A	Assembly and Crystal Mounting	64
		(1) Diamond mounting and alignment	64
		(2) Inserting the heater	65
		(3) Crystal mounting	66
	B	Calibration	68
		(1) Difficulties in calibration	68
		(2) Calibration with a thermocouple plus internal standard	68
		(3) Double internal x-ray standard	70
	C	X-ray Photography and Diffractometry	72
		(1) Modified goniometer head and heater wire supports	72
		(2) Precession photography	73
		(3) Diffractometry and data analysis	74
IV		Future Prospects	75

I. HEATING THE DIAMOND CELL

The previous chapters have dealt with crystallographic studies at high temperature or high pressure alone. Pressure and temperature must be combined, however, if comprehensive equation of state data are to be obtained. Crystallography at high temperature and pressure presents formidable challenges, both in design and calibration, and while many of these experimental difficulties have been resolved using a combination of procedures described in Chapters Two and

Three, this technology is still in the developmental stage. What follows, therefore, is largely a state-of-the-art presentation.

A. Early work

An obvious solution to the problem of subjecting a single crystal to simultaneous high temperature and high pressure is to heat the diamond anvil cell. Consequently, diamond cells were heated almost from the beginning of their development. Many of these early experimental techniques are not well documented, however, because heating of tungsten carbide anvil cells was routine in the late fifties and early sixties, and little design modification was required in heating diamond anvil devices. The unique experimental opportunities of optically transparent anvils did lead to some novel and effective heating techniques, however.

Perhaps the first, and certaintly the simplest, heating procedure of a diamond cell was used at the United States National Bureau of Standards, where a cell with pressurized sample was placed on a hot plate and warmed to a maximum temperature of more than 200°C. The cell was then transferred to a microscope where the high-temperature-pressure behaviour of the sample was observed during cooling (Van Valkenburg, personal communication).

Resistance heaters, with wire windings placed around the diamond and sample region, have been widely used to provide uniform temperatures for long time periods. These heaters, however, are limited to temperatures no greater than 1000°C owing to graphitization of the diamonds. Furthermore, gasketing materials now in use become too weak to support high pressures at temperatures above 800°C, placing an additional restriction on the temperature range. Yet another constraint is that an inert atmosphere is required above 700°C to avoid oxidation of diamonds.

A variety of resistance furnace configurations have been attempted. Several designs incorporate windings of from 2 to 4 cm in diameter, which heat both diamonds and supports (Van Valkenburg, 1962; Bassett and Takahashi, 1965; Fourme, 1968; Barnett, Block and Piermarini, 1973). In these diamond cells the sample is heated while the pressure loading mechanism remains relatively cool. Bassett and Ming (1972) used a novel long and narrow cell that fit into a standard tube furnace. A spring remained cool outside the furnace, while applying pressure to the diamond anvils via a long piston. Recent designs by Sung (1976) and Hazen and Finger (1979) incorporate a miniature winding 4 to 6 mm in diameter that heats the diamond and sample area of a small cell (see below).

Miyano and Sueno (1981) have constructed a diamond-anvil cell that is 3.2 cm in diameter, with a band or strip heater wrapped around the outside. They have reported a maximum temperature of 550°C, which is higher than that obtained in other single-crystal x-ray pressure cells. However, Miyano and Sueno report a loss of pressure above 300°C, probably due to heating and expansion of the load-bearing portions of the cell.

Another resistance heating technique, which has met with only limited success, is the use of the gasket or an internal strip heater as the resistance element (Moore *et al.*, 1970). These devices have not been used above 300°C owing to the rapid conduction of heat away from the sample chamber by diamond, which is one of the most efficient thermal conductors. Because of this property of diamond, successful resistance heating generally requires uniform heating of the anvils as well as the sample.

Lasers provide a method of heating samples to temperatures significantly greater than the decomposition point of diamond. Bassett and Ming (1972) and Ming and Bassett (1974) achieved steady state temperatures to 2000°C and pulsed temperatures of 3000°C in polycrystalline samples at pressures greater than 200 kbar with a YAG laser. High temperatures are attainable because a sample that absorbs laser energy, or is mixed with an absorber, may be heated to several thousand degrees centigrade, while the transparent diamonds remain relatively cool owing to their high thermal conductivity. An obvious difficulty of this technique is that large and nonuniform thermal gradients are produced in the sample chamber making calibration of temperature and pressure difficult. In addition, although lasers are ideal for heating polycrystalline samples in experiments where the diamond cell orientation is constant, cell orientation in single-crystal x-ray experiments must be varied, rendering uniform heating of a laser-irradiated sample difficult. For these reasons laser heating has not yet been adopted in single-crystal diamond-cell experiments.

B. Heating single crystals in the diamond cell

The great majority of high-temperature, diamond-cell experiments have been studies of polycrystalline samples. The few single-crystal studies have all employed small resistance furnaces to achieve high temperature. Weir *et al.* (1969) at the National Bureau of Standards employed a gasketed lever-arm cell (Figure 3-5) with resistance heater in studies of the structure and phase equilibria of crystallized fluids. Van Valkenburg has continued to use this cell, which has a maximum temperature range of about 300°C, in his studies of the solubility of crystals as a function of temperature and pressure (Baur and Van Valkenburg, 1979). Fourme's (1968) single-crystal diamond cell for use on an x-ray precession camera (Figure 3-9) was equipped with a resistance heater capable of sustaining 250°C. This device was employed by Andre *et al.* (1971) in the first combined pressure–temperature single-crystal x-ray experiments, in which the molecular arrangement of monochlorobenzene at 120°C and 14 kbar was deduced from x-ray photographic measurements.

Several of the single-crystal cells described in Chapter Three might be easily modified for high-temperature work by the addition of a resistance heater around the diamond and sample region. Designs by Keller and Holzapfel (1977, Figure 3-10), Schiferl (1977, Figure 3-7) and Merrill and Bassett (1974, Figure 3-6, see below) are especially easy to modify. Hazen and Finger (1979) added a heater to the Merrill and Bassett diamond cell, and produced the first three-dimensional, high-temperature, high-pressure crystal structure refinement in the summer of 1979. The construction and operation of this heated cell are described below.

II. DESIGN OF A HIGH-TEMPERATURE DIAMOND CELL

The Merrill and Bassett (1974) miniature diamond cell, as described in Chapter Three, has been modified for use at high temperatures to 600°C by the addition of a miniature resistance furnace (Hazen and Finger, 1981a). This section, derived from Hazen and Finger (1981a), includes details of the construction of the cell. Diamond anvils and triangular steel supports are identical to those described in Chapter Three. Other components are modified as follows.

A.   Diamond supports

Beryllium softens at temperatures above 200°C to the extent that some other backing material for diamonds is necessary. The additional requirement that the backing be relatively transparent to x-rays severely limits the selection. One satisfactory material is a hot-pressed boron carbide, 'Norbide' (Norton Corporation), which is both strong and has a low absorption coefficient in molybdenum $K\alpha$ radiation. Norbide is extremely hard and therefore must be machined by diamond wheel grinding. Drilling tends to weaken the material because the tensile strength of this boron carbide is lower than that of beryllium metal, and diamond support disc failure is a logical concern. For these reasons the support discs are machined without the optical access holes illustrated in Figure 3-14. The lack of visual access to the sample is one disadvantage of the present design. The external dimensions of the discs are 1.195 cm diameter and 0.295 cm thick.

B.   Gaskets

Inconel 750X (International Nickel Corporation) is used in all our high-temperature diamond-cell experiments because it retains its strength to approximately 800°C. Space requirements of the miniature resistance furnace (see below) restrict the size of the gasket to a disc only 3 mm in diameter with the sample chamber hole drilled in the centre. The dimensions of the hole are selected as described in Chapter Three.

C.   The miniature heater

The miniature resistance heater employs the design of Ohashi and Hadidiacos (1976), in which the furnace winding is composed of half platinum and half $Pt_{90}Rh_{10}$ thermocouple wire, with the join bead at the centre of the winding. Power is supplied to the furnace during half a cycle of an a.c. current, and the e.m.f. of the thermocouple is sensed during the other half and is used to regulate the power to the heater. A block diagram of this controller is illustrated in Figure 4-1 and the circuit is shown in Figure 4-2. A major advantage of this design is that only two wires are needed to supply power to the furnace and to sense the winding temperature.

The furnace winding consists of three to five loops of 0.038 cm diameter wire formed using the thread of a 10-32 screw as a template (Figure 4-3). The diameter of the winding is thus approximately 4 mm, and is large enough to fit

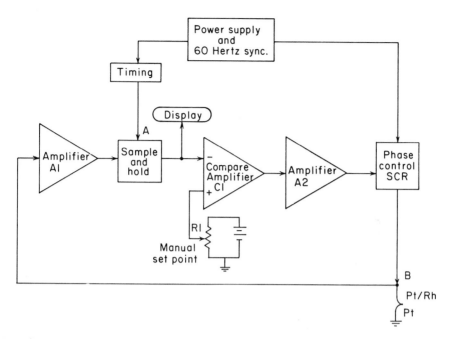

Figure 4-1. Schematic diagram of the microheater control circuitry. (From Ohashi and Hadidiacos, 1976, reproduced by permission of the Carnegie Institution of Washington)

around the diamonds and gasketed sample, between the two boron carbide supports. The winding fits inside the inner pyrophyllite insulating ring and the winding assembly is coated with Sauereisen high-temperature furnace cement (Sauereisen Cements Co.). The wire is doubled near the heater coil to concentrate heat in the sample region. Four-hole ceramic tubing is used to insulate the heater wire between the triangular steel pieces. Each wire is redoubled outside the cell to localize heating and reduce power requirements. The entire heater assembly (Figure 4-4) is easily removed, and the diamond cell may be operated as a standard high-pressure cell in its absence.

D. Insulation

Several types of thermal insulation have been employed in the early versions of the high-temperature Merrill and Bassett diamond cell. Two outer pyrophyllite insulating rings are machined to fit snugly around the boron carbide discs, between the discs and the three diamond alignment screws. These rings (Figure 4-5) have a small semi-cylindrical groove cut to fit the two-wire ceramic tube which leads from the heater. The pyrophyllite is heat-treated at 800°C for 2 hours to harden the rings partially.

Insulating washers of sheet mica are placed between the triangular pieces and boron carbide discs. The washers' outer diameter is 1.27 cm, inner diameter is 0.76 cm and the thickness is 0.05 cm. Thermally reflecting platinum foil is placed

62

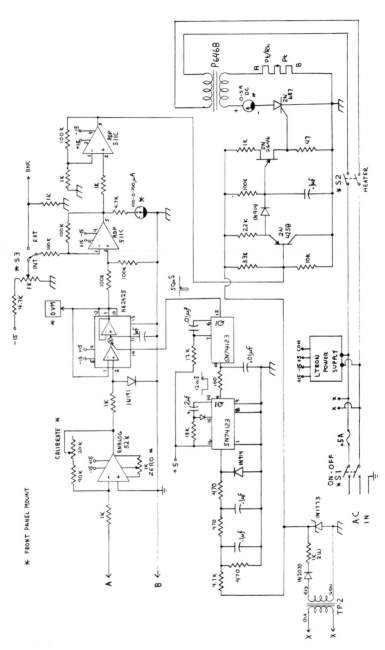

Figure 4-2. Circuit diagram for the microheater temperature controller, which employs a thermocouple as both heating and sensing element. Manufacturers are cited for reference only. (From Ohashi and Hadidiacos, 1976, reproduced by permission of the Carnegie Institution of Washington)

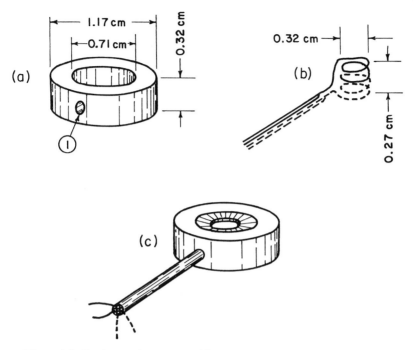

Figure 4-3. Resistance heater assembly.
A. The pyrophyllite support; (1) is a hole for 1.6 mm ceramic 4-hole tubing.
B. The winding is made of 0.33 mm diameter wire, with 3 or 4 turns at approximately 18 turns per cm. Solid line is Pt, dashed line is $Pt_{90}Rh_{10}$.
C. The assembled heater; Sauereisen high-temperature cement (Sauereisen Cements Company) holds the winding in place.
(From Hazen and Finger, 1981a, reproduced by permission of the American Institute of Physics)

Figure 4-4. Miniature furnace mounted in a diamond-anvil pressure cell

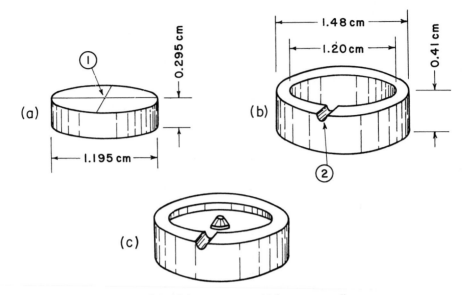

Figure 4-5. Components of the high-temperature, high-pressure cell.
A. Boron carbide disc; (1) is a small scribe mark for ease of diamond alignment.
B. Pyrophyllite insulating ring with groove (2) for ceramic tubing.
C. Diamond support assembly, consisting of A, B and diamond anvil.
(From Hazen and Finger, 1981a, reproduced by permission of the American Institute of Physics)

around all furnace parts, including washer-shaped sheets between the boron carbide discs and the furnace, a ring of Pt foil around the outside of the inner pyrophyllite ring, and a strip around the two-wire ceramic tube. Additional insulation is provided by Sauereisen cement, applied during cell assembly as described below.

III. OPERATION OF THE HIGH-TEMPERATURE DIAMOND CELL

A. Assembly and crystal mounting

(1) Diamond mounting and alignment

An exploded view of the high-temperature diamond cell is presented in Figure 4-6 to aid in assembly. Diamonds are glued to the boron carbide discs using epoxy cement, as in room temperature experiments. The diamonds commonly break loose during unloading owing to alteration of the cement at high temperature. Sauereisen high-temperature cement was not found to be effective because of poor adhesion to diamond. Alternative adhesives for the diamonds in these experiments would be desirable.

Boron carbide discs with diamonds attached fit snugly into the outer pyrophyllite insulating rings. These two components are bonded together with

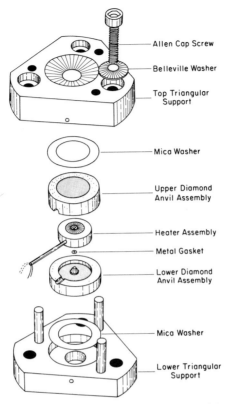

Figure 4-6. Exploded view of the high-temperature, high-pressure diamond-anvil cell. (From Hazen and Finger, 1981a, reproduced by permission of the American Institute of Physics)

Sauereisen cement. Mica insulating washers are placed into wells in the steel triangular pieces, and the two diamond anvil/boron carbide-pyrophyllite ring assemblies are then centred and aligned as described in Chapter Three, section III-A-(2). The orientation of the two outer pyrophyllite rings must be matched so that the two semi-cylindrical grooves are opposite each other, and are directed towards the top of the diamond cell. Once aligned and secured by the three small set screws, Sauereisen cement may be applied to spaces between the outer pyrophyllite discs and triangular pieces to provide additional insulation.

(2) Inserting the heater

The heater is inserted after diamond alignment. The inner pyrophyllite ring is designed to fit snugly inside the lip of the outer pyrophyllite ring, with the four-wire ceramic tube directed out the top side of the cell (Figure 4-4). Once in place, the furnace can be anchored with more Sauereisen cement. The windings of the heater should be concentric about the diamond anvils, but should not touch the anvils.

(3) Crystal mounting

The disc-shaped gasket is too small to be supported by jeweller's wax on the lower diamond. A small spot of epoxy cement or a light smear of vacuum grease around the diamond anvil edge has been found to be adequate to hold the gasket in position during gasket centring (Chapter Three, section III-A-(3)) and crystal mounting (Figure 4-7). Crystals are mounted as described in Chapter Three, III-A-(4). Ruby calibration crystals are not required owing to lack of optical access in the high-temperature cell. An x-ray calibration crystal will be included in most experiments, as described below.

Figure 4-7. Gasket assembly for the high-temperature, high-pressure cell. The circular gasket is approximately 3 mm in diameter and rests on the lower diamond anvil face

Figure 4-8. Assembled high-temperature, high-pressure
diamond-anvil cell on the four-circle diffractometer

Pressure fluids are used in the same way as described in Chapter Three, section
III-A-(5); however, methanol–ethanol may not be the best high-temperature
hydrostatic (i.e. isotropic with no strain) fluid. It is not possible to monitor the formation of bubbles in the gasket chamber, and the alcohol mixture is more likely to
leak out of the gasket hole. Liquids of greater viscosity, such as glycerin, index-of-
refraction oil, or even water, may be less likely to leak at high temperature and
low pressure. More experimentation is required to identify the best high-
temperature pressure media.

Once the crystal is mounted and fluid added, the diamond cell is carefully
closed, making sure that the ceramic tube fits into the hole defined by the two
outer pyrophyllite rings. The cell is now ready for pressurization. In order to keep
the pressure loading mechanism somewhat independent of the temperature of the
cell and thermal expansion of the screws, from two to four Belleville washers are
inserted with each of the three load screws (Figure 4-6). These screws are
tightened consecutively in 5° to 10° increments. The completely assembled high-
temperature diamond cell is illustrated in Figure 4-8.

B. Calibration

(1) Difficulties in calibration

One of the most difficult problems confronting researchers using the high-temperature diamond cell is calibration. The ruby fluorescence lines used to calibrate the diamond cell at high pressure become diffuse and therefore unsatisfactory for calibration above 200°C, although it has been suggested that line broadening and position might provide a measure of temperature and pressure (Barnett *et al.*, 1973). Furthermore, in many high-temperature diamond-cell experiments, including those with the miniature cell described above, lack of optical access precludes the use of spectral measurements for calibration.

It should be emphasized that it is not generally possible to calibrate pressure prior to heating the diamond cell, because the pressure inside the sample chamber is greatly influenced by differential expansion of the hydrostatic fluid medium compared with other components of the cell. In the modified Merrill and Bassett cell presented above, for example, an increase in temperature of the cell causes an increase in pressure of the sample when Belleville washers are present, but a decrease in pressure with no washers.

(2) Calibration with a thermocouple plus internal standard

Perhaps the most promising calibration procedure is a combination of a thermocouple outside the gasketed sample chamber and an internal cubic x-ray standard (Hazen and Finger, 1981b). In the miniature high-temperature cell the winding itself acts as a temperature-controlling thermocouple. This thermocouple is calibrated versus sample temperature by measuring the thermal expansion of a cubic standard material at room pressure in the fully assembled cell. It is assumed that for a given furnace winding the correlation between e.m.f. of the thermocouple and sample temperature is constant, because the diamond–gasket–sample configuration is very similar in all experiments. This temperature measurement is precise to ± 2°C over the course of a 24 hour experiment. The accuracy is reported as only ± 10°C, however, owing to uncertainties in the effects of small changes in experimental configuration as the gasket is crushed during changes in pressure.

Once temperature is determined using a thermocouple, the pressure can be deduced from the unit-cell dimensions of an internal pressure standard. A single-crystal x-ray pressure standard should have the following characteristics:

1. chemical and structural stability over the range of pressure-temperature conditions to be studied
2. no reactivity with fluid pressure medium
3. significant unit-cell variation with pressure for calibration sensitivity
4. reversibility (i.e. no hysteresis) in unit-cell dimensions after application of temperature and pressure

5. cubic symmetry for ease of identification of reflections and minimum interference with sample crystal
6. strong diffraction intensity to minimize volume of the standard
7. excellent cleavage for use of thin plate crystals that will uniformly shadow the sample without taking too much space
8. a well-known pressure–temperature–volume equation of state.

Few compounds meet these requirements. Alkali halides, such as NaCl, meet criteria 3–8, but they dissolve at elevated temperature in fluid pressure media. Rock salt-structure oxides, such as MgO, are relatively incompressible and thus do not meet criterion 3. One common substance that does satisfy all eight requirements is calcium fluoride, CaF_2 (the mineral fluorite), which is thus proposed as an internal, single-crystal, x-ray standard for high-temperature, high-pressure experiments.

Calcium fluoride has been used as an internal pressure standard in the first high-temperature, high-pressure, single-crystal x-ray experiments at the Geophysical Laboratory (Hazen and Finger, 1981b). Fluorite has perfect (111) cleavage, and plates 15 to 30 μm thick, and several hundred μm in diameter are easily produced by fracturing natural crystals. (Very pure colourless fluorite crystals are commonly available.) The thin flat plates may be cleaved with a razor into triangular or hexagonal forms. Triangular plates 150 μm on each of three edges and 15 μm thick were used in the first high-temperature, high-pressure studies. The standard crystal was attached to the lower diamond-anvil face with a small dab of silicone vacuum grease, and Cargille index-of-refraction oil ($n = 1.515$) was used as the hydrostatic pressure fluid. The flat plate remained within $\pm 1°$ of the diamond anvil surface throughout the experiments.

Unit-cell dimensions of CaF_2 were determined as described in Chapter 3, section III-D. If temperature and cell volume are known then pressure may be calculated from the $T–P–V$ equation of state of CaF_2. Thermal expansion, α, of CaF_2 (at 1 bar) has been reported by Sharma (1951), Batchelder and Simmons (1964) and Larionov and Malkin (1975). Their data yield the relationship (T in °C):

$$\alpha_{T,0} = 1/V(dV/dT)_P$$
$$= 5.553 \times 10^{-5} + 4.443 \times 10^{-8}T + 6.456 \times 10^{-11}T^2 \quad (4\text{-}1)$$

This equation is valid between 0° and 650°C.

Compressibility of CaF_2 (at 0°C) has been reported by several authors, including Wong and Schuele (1967) and Ho and Ruoff (1967). Below 40 kbar the compressibility of CaF_2 is approximately (P in Mbar):

$$\beta_{0,P} = -1/V(dV/dP)_T = 1.217 - 4.5P \quad (4\text{-}2)$$

The effect of temperature on β has been determined by Wong and Schuele (1968) and Nikanorov, Kardashev and Kas'kovich (1968):

$$\beta_{T,0} = \beta_{0,0}(1 + 0.00035T) \quad (4\text{-}3)$$

This formula is valid to at least 500°C.

The total volume change between ambient conditions and high T and P, represented by $V_{T,P}/V_0$, may be calculated as the sum of volume changes from ambient to T and room pressure, plus the volume change from high T to high T and P:

$$\frac{V_{T,P} - V_0}{V_0} = \frac{V_{T,0} - V_0}{V_0} + \frac{V_{T,P} - V_{T,0}}{V_0}$$

$$= \frac{\Delta V_1}{V_0} + \frac{\Delta V_2}{V_0}$$

The terms V_1 and V_2 may be calculated by integration of equations (4-1), (4-2) and (4-3):

$$\frac{\Delta V_1}{V_0} = \int_0^T \alpha_{T,0} dT$$

$$= 5.553 \times 10^{-5}T + 2.222 \times 10^{-8}T^2 + 2.152 \times 10^{-11}T^3 \quad (4\text{-}4)$$

and,

$$\frac{\Delta V_1}{V_0} = -\int_0^P \alpha_{T,P} dP \approx -\int_0^P (\beta_{T,0}\beta_{0,P}/\beta_0) dP$$

$$= -1.217P - 0.00043TP + 2.25P^2 + 0.00078TP^2 \quad (4\text{-}5)$$

Equations (4-4) and (4-5) represent a T–P–V equation of state for CaF_2 that is valid for the range 0° to 500°C and 0 to 0.04 Mbar. The functional form of this equation of state has *no* theoretical significance; it is an empirical fit to data from several previous experimental studies. The equations are well-suited to the determination of P, if T and V are known, however. Equations (4-4) and (4-5) are most conveniently used in graphical form (Figure 4-9), which is the pressure calibration function used in our crystallographic studies.

(3) Double internal x-ray standard

A calibration procedure that employs a double x-ray standard might be both highly precise and accurate, without relying on a thermocouple for temperature measurement. If two crystallographic directions can be found that differ greatly in their ratios of thermal expansion to compressibility (α/β), then the measurement of those two directions together uniquely defines both temperature and pressure. Consider, for example, the a and c axes of $CaCO_3$-I, the common mineral calcite, in its stable phase region at temperatures below 1000°C and pressures below 12 kbar. With increased temperature the c axis expands but the a axis contracts; with increased pressure both axes compress (i.e. $\alpha/\beta > 0$ for c and <0 for a). Figure

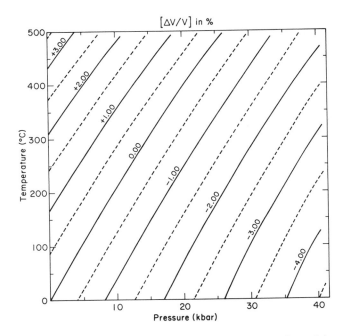

[ΔV/V] in %

Temperature (°C)

Pressure (kbar)

Figure 4-9. Pressure–temperature–volume nomogram for calcium fluoride. If temperature is known then the pressure is uniquely defined by the measured unit-cell volume. This relationship is the basis for the CaF$_2$ calibration system. (Reproduced from Hazen and Finger, 1981b, by permission of the International Union of Crystallography)

4-10 is a nomogram of calcite unit-cell lengths versus temperature and pressure. Unit-cell dimensions can be measured to at least 1 part in 10^4 using procedures outlined in Chapter Three. The resulting precision of temperature and pressure calibration, which are potentially the accuracy if the equation of state of CaCO$_3$-I is well known, are $\pm 5\,°C$ and ± 0.3 kbar. An obvious advantage of this procedure is that the calibration is tied directly to the reproducible equation of state of a standard material, rather than to the non-reproducible characteristics of a specific thermocouple.

In practice, an oriented flat (010) plate of calcite or some other uniaxial standard could be cut and then mounted in the pressure cell. The two axial distances of interest would be measured routinely as part of each high-temperature diamond cell experiment. Alternatively, two different cubic standards with different (α/β) could be employed. The ideal 'double x-ray standard' will depend on the range of temperature and pressure of each experiment.

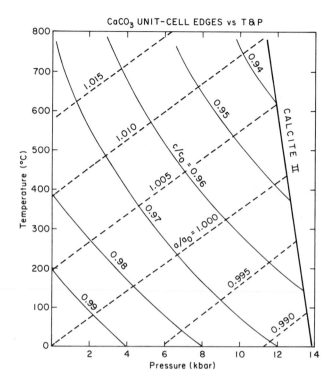

Figure 4-10. Pressure–temperature–cell constant nomogram for calcium carbonate. Calcite, $CaCO_3$, is hexagonal with differing pressure and temperature response of the a and c unit-cell axes. If a and c are known then temperature and pressure are uniquely defined. This relationship is the basis for the double standard calibration system

C. X-ray photography and diffractometry

* * * WARNING * * *

EXTREME CAUTION should be exercised while using the diamond cell in an x-ray beam because of the high level of scattered radiation, and the possibility of strongly directed diffraction from the diamond single crystals.

* * * ! ! ! * * *

(1) Modified goniometer head and heater wire supports

The diamond cell must be thermally insulated from the goniometer head to reduce heating and consequent damage to the mechanisms of the x-ray equipment. A pressure cell mounting cradle, identical to that illustrated in Figure 3-21 but constructed of pyrophyllite, accomplishes this objective. After machining, the pyrophyllite bracket is heat-treated at 1100°C for at least 8 hours to produce a hard, durable mount.

The heater wires, which exit from the top of the diamond cell, must be supported to prevent contact of the two leads, or wire fatigue and breakage. A simple mounting bracket has been constructed from 1/16 in sheet steel. This bracket attaches to the bottom triangular piece with two 10-32 screws, which fit into the back of the load-screw holes. The modified goniometer head and heater wire supports are illustrated in Figure 4-8, which shows the fully assembled high-temperature, miniature diamond cell.

(2) Precession photography

The high-temperature pressure cell may be operated in a manner similar to that described in Chapter Three, section III-C. Lacking an optical centring hole, the cell must be centred in the plane perpendicular to the x-ray beam using the x-rays and a fluorescent screen. With precession angle at zero, adjust the x and y goniometer head translations until maximum brightness is observed on the screen. This presumably corresponds to the x-ray beam passing through the gasket hole

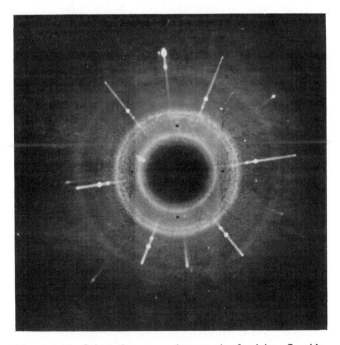

Figure 4-11. Orientation x-ray photograph of calcium fluoride, CaF_2, at 200°C and 10 kbar. This (111) photograph was taken with Polaroid Type 57 film, Mo radiation, 30 minute exposure at 45 kV and 15 ma. (Reproduced from Hazen and Finger, 1981b, by permission of the International Union of Crystallography)

and sample. Centring the cell parallel to the x-ray beam is accomplished as described previously. A precession orientation photograph of CaF_2 at high temperature and pressure is illustrated in Figure 4-11.

(3) Diffractometry and data analysis

There are two major differences in diffractometry and data analysis between the high-temperature diamond cell and room-temperature cell (see Chapter Three, III-D). These differences are a modified absorption function, due to boron carbide rather than beryllium diamond support discs, and the geometrical restrictions imposed by heater wires.

Absorption of boron carbide is handled in the same fashion as the beryllium absorption. However, because there is no hole in the boron carbide disc, the

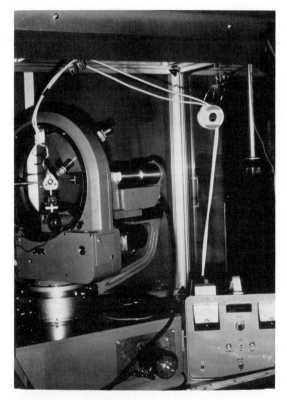

Figure 4-12. High-temperature, high-pressure diamond-anvil cell on the 4-circle diffractometer. The temperature controller is at the lower right. Power lines from the controller to the cell are routed along the top of the x-ray protective enclosure and down to the cell

absorption profile is more regular and may be approximated by an exponential function, as would be expected for a flat plate.

The operation of the diffractometer with the high-temperature cell mounted is restricted by the heater wires, which must remain clear of the x-ray beam. At the Geophysical Laboratory the power leads are routed from the top of the diffractometer to the roof of the protective enclosure (see Figures 4-8 and 4-12). A pulley and weight system is used to supply tension to the wires. As the χ-circle is rotated, posts on the instrument keep the wires from interfering with the beam or the collimators. An additional restriction is that the wires must not be twisted more than $270°$ with respect to the pressure cell. Such twisting is not a problem during data collection, because the cell is operated in the fixed-ϕ mode. However, during crystal centring using the 8-reflection method, normal techniques have been modified to restrict ϕ between $0°$ and $-180°$, and to minimize the number of times that ϕ is changed. In addition, χ is restricted to the range $\pm 180°$ and neither χ nor ϕ is permitted to be circular (i.e. $-180°$ is not equivalent to $+180°$).

In data collection, omega scans are performed as described in Chapter Three, and the data are corrected for diamond-cell absorption by the on-line computer.

IV. FUTURE PROSPECTS

Crystallography at combined high temperature and pressure holds great promise for the description and interpretation of crystalline behaviour at non-ambient conditions. Structure refinements have been completed on only two compounds, CaF_2 and $CaMgSi_2O_6$. High-temperature, high-pressure crystallography is thus a new and unexplored direction for the solid-state researcher.

Many instrumental improvements may be anticipated in the coming years. Extension of the pressure and temperature ranges and improved calibration procedures will enable the detection of ever more subtle structural changes with temperature and pressure. The use of different gasketing materials and pressure media may result in larger sample volumes and improved intensity data. Improved insulation may enhance the efficiency and stability of the system. All aspects of the experimental procedures should be examined for ways to increase the flexibility of the high-temperature high-pressure cell.

REFERENCES

Andre, D., R. Fourme and M. Renaud (1971) Structure cristalline du monochlorobenzène à 393 K et 14.2 kbars: Un affinement par group rigide. *Acta Cryst.* **B27**, 2371–2380.

Barnett, J. D., S. Block and G. J. Piermarini (1973) An optical fluorescence system for quantitative pressure measurement in the diamond anvil cell. *Rev. Scien. Instrum.* **44**, 1–9.

Bassett, W. A., and L.-C. Ming (1972) Disproportionation of Fe_2SiO_4 to $2FeO+SiO_2$ at pressures up to 250 kbar and temperatures up to 3000°C. *Phys. Earth Planet. Interiors* **6**, 154–160.

Bassett, W. A., and T. Takahashi (1965) Silver iodide polymorphs. *Amer. Mineral.* **50**, 1576–1594.

76

Batchelder, D. N., and R. O. Simmons (1964) Lattice constants and thermal expansivities of silicon and of calcium fluoride between 6° and 322°K. *J. Chem. Phys.* **41**, 2324–2329.

Baur, P., and A. Van Valkenburg (1979) Solution of a new high-pressure phase in $Ca(OH)_2 \cdot H_2O$. *Carnegie Inst. Washington Year Book* **78**, 606–608.

Fourme, R. (1968) Appareillage pour études radiocristallographiques sous pression et à température variable. *J. Appl. Cryst.* **1**, 23–30.

Hazen, R. M., and L. W. Finger (1979) A high-temperature diamond pressure cell for single-crystal studies. *Carnegie Inst. Washington Year Book* **78**, 658–659.

Hazen, R. M., and L. W. Finger (1981a) High-temperature diamond-anvil pressure cell for single-crystal studies. *Rev. Scien. Instrum.* **52**, 75–79.

Hazen, R. M., and L. W. Finger (1981b) Calcium fluoride as an internal pressure standard in high-pressure/high-temperature crystallography. *J. Appl. Cryst.* **14**, 234–236.

Ho, P. S., and A. L. Ruoff (1967) Pressure dependence of the elastic constants and an experimental equation of state for CaF_2. *Phys. Rev.* **161**, 864–869.

Keller, R., and W. B. Holzapfel (1977) Diamond anvil device for x-ray diffraction on single crystals under pressures up to 100 kbar. *Rev. Scien. Instrum.* **48**, 517–523.

Larionov, A. L., and B. Z. Malkin (1975) Thermal expansion of calcium fluoride. *Phys. Stat. Sol.* **B68**, K103–K105.

Merrill, L., and W. A. Bassett (1974) Miniature diamond anvil pressure cell for single crystal x-ray diffraction studies. *Rev. Scien. Instrum.* **45**, 290–294.

Ming, L.-C., and W. A. Bassett (1974) Laser heating in the diamond anvil press up to 2000°C sustained and 3000°C pulsed at pressures up to 260 kbar. *Rev. Scien. Instrum.* **45**, 1115–1118.

Miyano, S., and S. Sueno (1981) High temperature diamond anvil cell for single crystal x-ray experiments. *Twelfth International Congress of Crystallography Collected Abstracts.* International Congress of Crystallography, Ottawa, Canada, C-318.

Moore, M. J., D. B. Sorensen and R. C. DeVries (1970) A simple heating device for diamond anvil high pressure cells. *Rev. Scien. Instrum.* **41**, 1665–1666.

Nikanorov, S. P., B. K. Kardashev and N. S. Kas'kovich (1968) Temperature dependence of elastic constants of calcium fluoride. *Soviet Phys. Sol. State* **10**, 703–705.

Ohashi, Y., and C. G. Hadidiacos (1976) A controllable thermocouple microheater for high-temperature microscopy. *Carnegie Inst. Washington Year Book* **75**, 828–832.

Schiferl, D. (1977) 50-kilobar gasketed diamond anvil cell for single-crystal x-ray diffractometer use with the crystal structure of Sb up to 26 kilobars as a test problem. *Rev. Scien. Instrum.* **48**, 24–30.

Sharma, S. S. (1951) Thermal expansion of calcium fluoride. *Proc. Indian Acad. Sci.* **31A**, 261–274.

Sung, C.-M. (1976) New modification of the diamond anvil press: a versatile apparatus for research at high pressure and high temperature. *Rev. Scien. Instrum.* **47**, 1343–1346.

Van Valkenburg, A. (1962) High-pressure microscopy. In: *High-Pressure Measurement*, A. A. Giardini and E. C. Lloyd, (eds.), Washington, DC: Butterworths, 87–94.

Weir, C. E., G. J. Piermarini and S. Block (1969) Instrumentation for single crystal x-ray diffraction at high pressures. *Rev. Scien. Instrum.* **40**, 1133–1136.

Wong, C., and D. E. Schuele (1967) The pressure derivatives of the elastic constants of CaF_2. *J. Phys. Chem. Solids* **28**, 1225–1231.

Wong, C., and D. E. Schuele (1968) Pressure and temperature derivatives of the elastic constants of CaF_2 and BaF_2. *J. Phys. Chem. Solids* **29**, 1309–1330.

Chapter 5

The Parameters of a Crystal Structure

CONTENTS

I	Characterizing and Comparing Crystal Structures	77
II	Propagation of Errors	78
III	Changes in the Unit Cell	79
	A Volume changes	79
	B Unit-Cell Variation and the Strain Ellipsoid	80
IV	Changes in Atomic Positions	82
	A Bond Distances	82
	(1) Thermal corrections to bond distances	82
	(2) Mean bond distance and error estimations	84
	B Polyhedral Volume	85
	C Polyhedral Distortions	86
V	Thermal Parameters	86
	A Harmonic Vibrations	86
	B Rigid-body Motion	87
	C Anharmonic Vibrations	87
VI	Conclusion to Part One	88

I. CHARACTERIZING AND COMPARING CRYSTAL STRUCTURES

A full and complete description of the structure of a crystal requires knowledge of the spatial and temporal distributions of all atoms in the crystal. By definition the crystal has periodicity, so the spatial terms can be represented by (1) the size and shape of the unit cell, (2) the space group of the atomic arrangement, and (3) the fractional coordinates of all symmetrically distinct atoms. A complete description of the temporal variation is impossible for all real materials and a simplifying assumption of independent atoms with harmonic vibrations is usually made. This assumption implies thermal ellipsoids of constant probability density, and these

77

ellipsoids constitute the fourth element of the structure description. The determination of these structural parameters has remained a major objective of crystallographers since the first x-ray studies of the early twentieth century.

Although the majority of structures can be characterized by these four elements alone, many atomic arrangements are more easily conceptualized with the aid of additional descriptors derived from the basic set. Among the numerous additional parameters introduced to describe atomic arrangements are interatomic or bond distances, bond angles, coordination numbers, thermal ellipsoids, polyhedral volumes, polyhedral distortions, and packing parameters. These quantitative measures of a crystal structure are not essential, but they are of great value in illustrating and comparing different structures.

Closely related structures, such as members of a solid solution or a specific compound at different temperatures or pressures, may be described with a number of *comparative* parameters. Comparative parameters add no new data to the crystal structure description, but they are invaluable in characterizing subtle changes in structure. The reader must be aware that many comparisons involve subtraction, explicit or implicit, of two quantities that have similar magnitudes. In such cases the error associated with the difference may become very large. It is essential to propagate errors in the initial parameters to the derived quantity being investigated. In this chapter, several comparative parameters are described, along with procedures for calculation of these quantities *and* the associated errors.

II. PROPAGATION OF ERRORS

When a derived quantity is computed, the effect of the input variable errors on the result is highly dependent upon the nature of the calculation being performed. In the case of the difference between two similar quantities, for example, the percentage error may become very large. This section contains formulations used to propagate error values.

If y is a computed value that is a function of experimentally determined variables $x_1, x_2, \ldots, x_n$ such that

$$y = f(x_1, x_2, \ldots, x_n) \tag{5-1}$$

then the computed variance of y, denoted σ_y^2 is found from

$$\sigma_y^2 = \sum_{j=1}^{n} \sum_{k=1}^{n} \left(\frac{\partial f}{\partial x_j}\right) \left(\frac{\partial f}{\partial x_k}\right) \text{cov}(x_j, x_k) \tag{5-2}$$

where cov (x_j, x_k) is the covariance between x_j and x_k, also written as σ_{ij}^2. In many cases the correlations between the variables are unknown and common practice is to assume such correlations are zero. Equation (5-2) then simplifies to

$$\sigma_y^2 \cong \sum_{j=1}^{n} \left(\frac{\partial f}{\partial x_j}\right)^2 \sigma_{xj}^2, \tag{5-3}$$

where σ^2_{xj} is the variance of x_j. If the functional form of y is simple enough, the required partial derivatives are computed directly. For complicated functions it may be necessary to calculate numerically with the approximation

$$\frac{\partial f}{\partial x_j} \cong \frac{f(x_1, x_2, \ldots, x_j + \Delta, \ldots, x_n) - f(x_1, x_2, \ldots, x_n)}{\Delta} \tag{5-4}$$

The formal definition of a derivative is attained from equation (5-4) by taking the limit as $\Delta \to 0$. In most calculations a satisfactory value for the derivative is obtained for $\Delta \cong \sigma_{xj}/2$.

III. CHANGES IN THE UNIT CELL

The most widely reported continuous structural variations with temperature, pressure or composition are those associated with the unit cell. The lattice can be determined without performing a complete three-dimensional structure refinement. Furthermore, powder x-ray diffraction is sufficient to measure the unit cell in most materials. There is, consequently, a large body of data on the variation of cell dimensions with intensive variables. Variations of the unit cell are commonly reported in terms of volume changes, linear changes or the strain ellipsoid, as described below.

A. Volume changes

Characterizing changes of unit-cell (or molar) volume with temperature, pressure and composition is a major objective of equation of state research. Hundreds of studies on temperature–pressure–composition–volume (T–P–X–V) systematics, combined with theoretical treatments of equations of state, have led to a great variety of proposed functional forms for molar volume changes with other intensive variables. Equation of state models will not be reviewed in great detail here, because single-crystal, x-ray studies are not usually designed to produce the most precise T–P–X–V data. (See, however, Part Two, Chapters Six, Seven and Eight.)

In an ideal crystal, molar volume is simply related to unit-cell volume by

$$V_{molar} = N_0 \times \frac{V_{unit\ cell}}{Z}$$

where Z is the number of formula units per unit cell, and N_0 is Avogadro's number (6.023×10^{23} mole^{-1}). In real crystals, however, dislocations, stacking faults, vacancies and other nonperiodic defects may cause molar volume to be slightly greater than that predicted by the unit-cell measurement (i.e. the measured density may be less than the calculated density). Nevertheless, x-ray determination of unit-cell volume provides the most precise means of measuring molar volume.

Changes of molar volume with T, P or X may be described by the coefficient of thermal expansion, α_v, the compressibility, β_v, and the 'coefficient of compositional expansion,' γ_v:

$$\alpha_V = 1/V(\partial V/\partial T)_{P,X} \qquad (5\text{-}5)$$

$$\beta_V = -1/V(\partial V/\partial P)_{T,X} \qquad (5\text{-}6)$$

$$\gamma_V = 1/V(\partial V/\partial X)_{T,P} \qquad (5\text{-}7)$$

These parameters are explored further in Chapters Six, Seven and Eight, respectively.

B. Unit-cell variation and the strain ellipsoid

Linear changes of the unit cell with temperature, pressure and composition are relatively easy to measure, and provide important information regarding structural changes with these variables. As temperature, pressure or composition is changed, a spherical element of the original crystal will, in general, deform to an ellipsoid. This ellipsoid must be aligned with the unit-cell axes in cubic, hexagonal, tetragonal or orthorhombic crystals because of symmetry constraints. Therefore, axial changes of the unit-cell edges completely define the dimensional variation of the lattice. In monoclinic and triclinic crystals, on the other hand, unit-cell angles may also vary. A cataloguing of changes in each axial direction does not, therefore, reveal all significant changes. In the triaxial strain ellipsoid, major and minor ellipsoid axes represent the orthogonal directions of maximum and minimum change in the crystal. Relationships between the strain ellipsoid and the crystal must be calculated as described by Ohashi and Burnham (1973).

The usefulness of the strain ellipsoid may be illustrated by considering the behaviour of the sodium–aluminum silicate, $NaAlSi_3O_8$ (the mineral albite), at high temperature. All three crystallographic axes of this triclinic mineral are observed to expand between room temperature and 900°C. Calculation of the strain ellipsoid, however, reveals that one principal direction actually contracts as the temperature is increased (Ohashi and Finger, 1973; see also the example in Appendix 2).

The strain ellipsoid may be derived from two related sets of unit-cell parameters as follows (after Ohashi and Burnham, 1973). Let a_i, b_i and c_i represent direct unit-cell vectors before ($i = 0$) and after ($i = 1$) a lattice deformation. A strain tensor [**S**] may be defined in terms of these vectors:

$$\mathbf{S} \cdot \mathbf{a}_0 = \mathbf{a}_1 - \mathbf{a}_0 \qquad (5\text{-}8)$$

or, postmultiplying by the reciprocal lattice vector, $\tilde{\mathbf{a}}_0^*$,

$$\mathbf{S} = \mathbf{a}_1 \cdot \tilde{\mathbf{a}}_0^* - \mathbf{I} \qquad (5\text{-}9)$$

where **I** is the unitary 3×3 matrix.

Lattice strain is more easily visualized in a Cartesian system, rather than the crystallographic system represented by [**S**]. Cartesian unit base vectors are defined with $z \parallel c$, $x \parallel a^*$ and $y \parallel z \times x$. Lattice vectors may be expressed in terms of x, y and z:

$$\mathbf{a}_i = \mathbf{Q}_i \mathbf{x}_i \quad (i = 0, 1) \tag{5-10}$$

and reciprocal lattice vectors are given by

$$\mathbf{a}_i^* = \mathbf{Q}_i^{-1} \mathbf{x}_i \quad (\text{Note } \mathbf{x}_i = \mathbf{x}_i^*) \tag{5-11}$$

The transformation matrix $\mathbf{Q}_i$ is

$$\mathbf{Q}_i = \begin{bmatrix} \dfrac{a_i p_i}{\sin \alpha_i} & \dfrac{a_i(\cos \gamma_i - \cos \alpha_i \cos \beta_i)}{\sin \alpha_i} & a_i \cos \beta_i \\ 0 & b_i \sin \alpha_i & b_i \cos \alpha_i \\ 0 & 0 & c_i \end{bmatrix} \tag{5-12}$$

where

$$p_i = (1 - \cos^2\alpha_i - \cos^2\beta_i - \cos^2\gamma_i + 2 \cos \alpha_i \cos \beta_i \cos \gamma_i)^{\frac{1}{2}}$$

Let $\mathbf{R}$ represent the pure-rotational transformation matrix from the $\mathbf{x}_0$ Cartesian system to the $\mathbf{x}_1$ Cartesian system:

$$\mathbf{x}_1 = \mathbf{R}\mathbf{x}_0 \tag{5-13}$$

Substituting equations (5-10), (5-11) and (5-13) into equation (5-9):

$$\begin{aligned} \mathbf{S} &= \mathbf{Q}_1\mathbf{x}_1(\widetilde{\tilde{\mathbf{Q}}_0^{-1}\mathbf{x}_0}) - \mathbf{I} \\ &= \mathbf{Q}_1\mathbf{R}\mathbf{x}_0\tilde{\mathbf{x}}_0\mathbf{Q}_0^{-1} - \mathbf{I} \\ &= \mathbf{Q}_1\mathbf{R}\mathbf{Q}_0^{-1} - \mathbf{I} \end{aligned} \tag{5-14}$$

The crystallographic strain tensor [S] is transformed into the Cartesian strain tensor [E] by a similarity transform:

$$\begin{aligned} \mathbf{E} &= \mathbf{Q}_0^{-1}\mathbf{S}\mathbf{Q}_0 \\ &= \mathbf{Q}_0^{-1}\mathbf{Q}_1\mathbf{R} - \mathbf{I} \end{aligned} \tag{5-15}$$

It is not possible to assess pure rotation between two sets of unit-cell data, but if we exclude pure rotation $\mathbf{R}$, then the *asymmetric* strain tensor is $\mathbf{E}_r = \mathbf{Q}_0^{-1}\mathbf{Q}_1 - \mathbf{I}$, and the *symmetric* strain tensor [ε] is given as:

$$\begin{aligned} \varepsilon &= 1/2(\mathbf{E}_R + \tilde{\mathbf{E}}_R) \\ &= 1/2(\mathbf{Q}_0^{-1}\mathbf{Q}_1 + \widetilde{\mathbf{Q}_0^{-1}\mathbf{Q}_1}) - \mathbf{I} \end{aligned} \tag{5-16}$$

The values of $\mathbf{Q}_0$ and $\mathbf{Q}_1$ may be calculated from equation (5-12), which combined with equation (5-16) yields ε.

The strain ellipsoid is usually presented in terms of the fractional change of major, minor and (orthogonal) intermediate strain axes per °C, or kbar, or mole per cent, combined with the angles between strain axes and crystallographic axes.

Errors in strain ellipsoid quantities are obtained by application of equation (5-2) if unit-cell covariances are known; otherwise equation (5-3) is used. All derivatives are obtained by incrementing the lattice constants in turn, recomputing the strain values and substituting into equation (5-4). A computer program for the calculation of strain ellipsoids from unit-cell data is presented in Appendix II at the end of Part One.

IV. CHANGES IN ATOMIC POSITIONS

Atomic positions are usually reported in terms of fractional coordinates. Changes in atomic fractional coordinates with temperature, pressure or composition, however, do not, in general, convey a clear picture of the nature of structural changes. Far more enlightening are changes in linear and volume subunits of a structure, analogous to variation of unit-cell edges and volume. These comparative parameters, including bond linear compression or expansivity, polyhedral volume compression or expansivity and changes in polyhedral distortion, are described below.

A. Bond distances

Interatomic distance is a function of temperature, pressure and composition. Bond distances have, therefore, been used routinely in describing structural variations with T, P and X. However, caution must be used in calculating distances and their errors, as described below.

(1) Thermal corrections to bond distances

In the simplest formalism bond distance may be calculated as the separation between atomic positions, as recorded in a crystal structure refinement. This 'uncorrected' bond distance, however, is valid only near absolute zero. At higher temperatures thermal vibrations of atoms result in average separation of atoms that may be significantly different from the distance between mean atomic positions (Busing and Levy, 1964). Consider the two-dimensional analogue, illustrated in Figure 5-1. If thermal vibrations of two atoms are highly correlated and parallel, then the mean separation will approximate the centroid separation. If, on the other hand, motion is anti-parallel and highly correlated then the mean separation will be greater than the centroid separation.

Consider the example of aluminum–oxygen bond distances in $NaAlSi_3O_8$, as described by Winter et al. (1977). Distances versus temperature for several types of thermal corrections are illustrated in Figure 5-2. The uncorrected Al–O distances, which actually decrease with temperature, are clearly underestimates of the average distance between atoms. More realistic models for Al–O distance variations with temperature are given by the lower limit or the riding motion examples, which represent highly correlated vibrations. Correlated vibrations occur in some strongly bonded atom pairs (e.g. some Si–O bonds), and riding motion is characteristic of hydrogen in OH groups. Uncorrected motion, in which

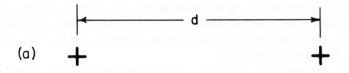

(a)

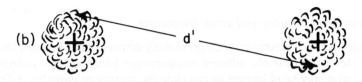

(b)

Figure 5-1. Two measures of interatomic distances.
A. Separation, d, of the time-averaged positions of two atoms.
B. Time-averaged separation, d', of two vibrating atoms. In general, if motions are non-correlated or anti-correlated, then d' will be greater than d

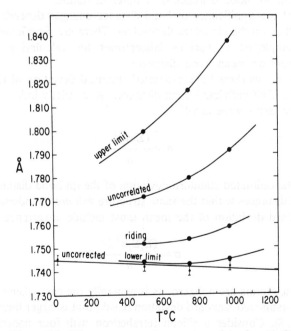

Figure 5-2. Distance versus temperature data for Al–O bonds in $NaAlSi_3O_8$ (the mineral albite) from Winter et al. (1977). The interpretation of Al–O bond behaviour versus temperature is critically dependent on the nature of the unknown vibration correlations. (Reproduced by permission of the Mineralogical Society of America)

atoms vibrate independently, is an intermediate case and may best represent metal–metal and anion–anion separations in oxides and silicates. Anti-correlated motion yields the upper bound for interatomic distances, and is an overestimate for most cation–anion separations.

In general, the nature of correlated atomic vibrations in crystals is unknown, and thermal corrections to bond distance cannot be made. Nevertheless, the possible differences between instantaneous and time-averaged atomic separations should be remembered when studying structures at high temperature.

(2)   Mean bond distance and error estimations

In structures with several crystallographically distinct bonds between the same types of atoms (for example, different metal–oxygen bonds within a coordination polyhedron) it may be of interest to calculate the average or *mean* bond distance, $\bar{d}$:

$$\bar{d} = \sum_{i}^{n} \frac{d_i}{n} \qquad (5\text{-}17)$$

where d_i is the ith bond distance of n different bonds.

Analysis of the significance of bond-distance changes depends on calculated errors as well as on the distances themselves. There are significant differences in procedures employed at various laboratories for calculating and reporting estimated errors on mean bond distances.

Most previous workers have estimated standard deviation of the mean bond distance, $\bar{\sigma}$, as if all individual bond distances were strictly independent (i.e. as if all covariance terms were zero):

$$\bar{\sigma} = \frac{(\sum_i \sigma_i^2)^{\frac{1}{2}}}{n} \qquad (5\text{-}18)$$

where σ_i is the estimated standard deviation of the ith bond distance. In general, several bond distances within the same structure will *not* be independent, and the correct standard deviation of the mean must include covariance terms, σ^2_{ij}:

$$\bar{\sigma} = \frac{(\sum_i \sum_j \sigma^2_{ij})^{\frac{1}{2}}}{n} \qquad (5\text{-}19)$$

If covariance terms are nonzero, as is generally the case, then equation (5-19) may yield an estimated standard deviation (e.s.d.) that is larger than that given by equation (5-18). Consider a silicon tetrahedron with four nonequivalent Si–O distances of 1.61, 1.62, 1.63 and 1.64 A (all with $\sigma = 0.01$ A). If all covariance terms are zero, then the mean distance (and e.s.d.) are 1.625 (0.005) A. In the other extreme, when all covariance terms are 1.0, the mean distance (and e.s.d.) are 1.625 (0.01) A.

In practice the covariance terms between nonequivalent calculated bond distances are not known, and equation (5-19) cannot be applied. A conservative approach in reporting estimated standard deviations of the mean bond distance is to assume all correlation coefficients are equal to one (the worst case). In equation (5-19) $\sigma_i\sigma_j$ may be substituted for σ^2_{ij} and the summation performed to yield:

$$\bar{\sigma} \leqslant \frac{\Sigma\sigma_i}{n} \tag{5-20}$$

It is thus evident that equation (5-20) gives an upper limit for the estimated standard deviation of a mean bond distance. Nevertheless, it is prudent to use this equation, rather than equation (5-18), which usually underestimates the mean error.

B. Polyhedral volume

In numerous compounds, including most of those characterized as 'ionic' by Pauling (1960), it is useful to examine cation coordination polyhedra as subunits of the structure. The volumes of these structural elements change with temperature, pressure and composition, and analysis of these changes is important in detailing the nature and origin of structural variations.

Polyhedral volume is defined as the space enclosed by passing planes through each set of three coordinating anions of a central cation. This volume may be calculated from unit-cell data and atomic coordinates. A computer program to calculate polyhedral volumes is listed in Appendix III, at the end of Part One. A similar program has been described by Swanson and Peterson (1980).

Polyhedral volumes may be used to calculate polyhedral thermal expansivity, compressibility and compositional expansivity in the same way as $T-V$, $P-V$, or $X-V$ data are used (equations 5-5, 5-6 and 5-7). In general, however, these expressions of approximate polyhedral volume change are also easily calculated from linear changes. For polyhedra that do not undergo severe distortion, the polyhedral thermal expansivity, compressibility and compositional expansivity are given by:

$$\alpha_{V\ polyhedral} \cong 3/\bar{d}(\partial d/\partial T) \tag{5-21}$$

$$\beta_{V\ polyhedral} \cong -3/\bar{d}(\partial d/\partial P) \tag{5-22}$$

$$\gamma_{V\ polyhedral} \cong 3/\bar{d}(\partial d/\partial X) \tag{5-23}$$

where $1/\bar{d}(\partial\bar{d}/\partial T)$ is mean linear thermal expansivity, etc. These relationships also allow the calculation of 'effective' polyhedral parameters for planar atomic groups. For example, if the mean C–O compressibility of a CO_3 group is known, then the effective polyhedral compressibility may be calculated from equation (5-22).

C. Polyhedral distortions

Cation polyhedra in most structures only approximate to regular geometrical forms. Deviations from regularity may be characterized in part by using distortion parameters. Two commonly used polyhedral distortion indices are quadratic elongation and bond angle variance, which are based on values of bond distances and bond angles, respectively (Robinson et al., 1971).

Quadratic elongation, $\langle\lambda\rangle$, is defined as:

$$\langle\lambda\rangle = \sum_{i=1}^{n} [(l_i/l_0)^2/n] \qquad (5\text{-}24)$$

where l_0 is the centre-to-vertex distance of a regular polyhedron of the same volume, n is the coordination number of the central atom, and l_i is the distance from the central atom to the ith coordinating atom. A regular polyhedron has a quadratic elongation of 1, whereas distorted polyhedra have values greater than 1.

Bond angle variance, σ^2, is defined as:

$$\sigma^2 = \sum_{i=1}^{n} [(\theta_i - \theta_0)^2/(n-1)] \qquad (5\text{-}25)$$

where θ_0 is the ideal bond angle for a regular polyhedron (e.g. 90° for an octahedron or 109.47° for a tetrahedron), n is the coordination number, and θ_i is the ith adjacent bond angle from outer, to central, to outer atoms. Angle variance is zero for a regular polyhedron and positive for a distorted polyhedron.

The computer program for calculating polyhedral volumes (Appendix III) may also be used to determine quadratic elongation and bond angle variance for octahedra and tetrahedra, and the errors associated with these quantities. To simplify the input, this program ignores covariances of atomic positional parameters and uses equation (5-3) to calculate errors. As a result the uncertainty in the bond angle variance is always calculated to be zero.

V. THERMAL PARAMETERS

A. Harmonic vibrations

The intensity of radiation scattered by a crystal depends not only on the types and relative positions of atoms, but also on the extent that each scattering atom deviates from a point. Such deviations may represent static configurations, as in the case of a positionally disordered site, or dynamic displacements that result from thermal motions. A number of methodologies have been devised to calculate such factors.

The simplest density distribution model is applicable to independent atoms, where all delocalization is due to harmonic thermal motion. The instantaneous position of the atom is described by a trivariate-normal distribution or three-dimensional Gaussian distribution (Johnson, 1970). This representation is the familiar anisotropic temperature factor, in which surfaces of equal probability are ellipsoids. When the ellipsoid is constrained to be a sphere, the isotropic temperature factor results. Hamilton (1959) gives the conversion between a given set of anisotropic coefficients and the equivalent isotropic temperature factor.

B. Rigid-body motion

A Gaussian distribution is not adequate to describe effects that arise from anharmonic motion of atoms. Furthermore, the Gaussian model fails for motion arising from rotation of a rigid multi-atom entity, such as a CO_3 group, about an axis. Techniques for dealing with such motion were presented by Shomaker and Trueblood (1968) for molecular crystals. In their approach the anisotropic temperature factor coefficients were refined as if the atoms were vibrating independently. These atomic coefficients were then used to calculate the coefficients of three tensors, **T**, **L** and **S**, which describe translation, libration and screw-coupling of the molecule, respectively.

Johnson (1970) developed a procedure for treatment of segmented rigid bodies. His procedure, however, relies on fitting the observed data (the intensities) to an approximation (the anisotropic thermal factors), and *then* molecular motions to these intermediate factors.

An alternative approach to determining molecular vibrations is to determine the effects of molecular motion on the *calculated* structure factors and to refine **T**, **L** and **S** tensor coefficients directly. Prince and Finger (1973) have reviewed the pertinent literature and present formulae for the direct refinement of rigid-body motion. These equations have been implemented by Finger and Prince (1975), who present a computer program for refinement of rigid-body parameters. The utility of this technique was demonstrated by Finger (1975) in his study of rigid-body motion of the carbonate ion in calcium and magnesium carbonates. It was possible to refine **T**, **L** and **S** tensor coefficients and to relate them to lattice mode vibration frequencies measured by infrared and Raman spectroscopy.

Although rigid-body rotations require inclusion of terms in the structure factor equation that appear to be anharmonic, the actual motions of the molecule may remain harmonic as long as amplitudes are small.

C. Anharmonic vibrations

Anharmonic contributions to structure factors have been extensively reviewed by Willis and Pryor (1975). For sites of high symmetry, factors relating thermal motion to the interatomic potential are written and used to modify the structure factor equations. Such equations are useful for predicting behaviour of anharmonic terms with temperature, but can be very difficult to derive. Johnson (1970) describes an especially convenient method for presenting anharmonic terms. In his procedure the probability density function (p.d.f.) includes cumulant tensors of progressively higher rank. The first four tensors describe the mean, the variance–covariance, the skewness and the kurtosis of the distribution. If the

Fourier transform of the p.d.f. is taken then a modified structure factor equation results. Although the cumulant coefficients are easily incorporated into a least-squares program (e.g. Finger and Prince, 1975), it is quite difficult to interpret the results, because there is no direct physical effect that contributes to each term. In addition, the number of parameters increases rapidly for atoms in general positions, for which there are 10 independent third-cumulant terms and 15 fourth-order coefficients.

Another model, called a quasi-orthogonal expansion for higher-order atomic thermal motion, is described by Johnson (1980). This model contains only one more parameter than conventional refinements. The additional term changes the structure factor equation from a series expansion at one limit to a Hermite polynomial at the other. The advantages of this model are that the numbers of parameters can be reduced and the convergence of the refinement is more stable.

All discussions of thermal factors must be tempered by the realization that the functional dependence of these terms is approximately an exponential to the negative second power of $\sin \theta$. Incorrect representation of other terms, such as absorption, scattering factors for x-rays or extinction, which have a similar form, will affect the refined parameters that represent 'thermal' motion.

VI. CONCLUSION TO PART ONE

The previous sections of this book include information on the attainment of nonambient conditions in an x-ray, single-crystal experiment and on the collection and analysis of data from such experiments. The procedures described in this state-of-the-art summary are not the final word in high-temperature and high-pressure crystallography, for the rapid progress of the past decade will no doubt continue. Each researcher is encouraged to try new materials, new technology and new procedures in the quest for higher temperatures and pressures and more precise data and analysis.

REFERENCES

Busing, W. R., and H. A. Levy (1964) The effect of thermal motion on the estimation of bond lengths from diffraction measurements. *Acta Crystallogr.* **17**, 142–146.

Finger, L. W. (1975) Least-squares refinement of the rigid-body motion parameters of CO_3 in calcite and magnesite and comparison with lattice vibrations. *Carnegie Inst. Washington Year Book* **74**, 572–575.

Finger, L. W., and E. Prince (1975) A system of Fortran IV computer programs for crystal structure calculations. *Natl. Bureau Standards (U.S.) Technical Note* 854.

Hamilton, W. C. (1959) On the isotropic temperature factor equivalent to a given anisotropic temperature factor. *Acta Crystallogr.* **12**, 609–610.

Johnson, C. K. (1970) Generalized treatments for thermal motion. In *Thermal Neutron Diffraction*, B. T. M. Willis (ed.), Oxford University Press.

Johnson, C. K. (1980) Thermal motion analysis. In *Computing in Crystallography*, R. Diamond, S. Ramaseshan and K. Venkatesan (eds.), Indian Academy of Sciences, Bangalore, India.

Ohashi, Y., and C. W. Burnham (1973) Clinopyroxene lattice deformations: The roles of chemical substitution and temperature. *Am. Mineral.* **58**, 843–849.

Ohashi, Y., and L. W. Finger (1973) Lattice deformation in feldspars. *Carnegie Inst. Washington Year Book* **72**, 569–573.

Pauling, L. (1960) *The Nature of the Chemical Bond*, Cornell Univ. Press, Ithaca, NY, 644 pp.

Prince, E., and L. W. Finger (1973) Use of constraints on thermal motion in structure refinement of molecules with librating side groups. *Acta Crystallogr.* **B29**, 179–183.

Robinson, K., G. V. Gibbs and P. H. Ribbe (1971) Quadratic elongation: a quantitative measure of distortion in coordination polyhedra. *Science* **172**, 567–570.

Shomaker, V., and K. N. Trueblood (1968) On the rigid-body motion of molecules in crystals. *Acta Crystallogr.* **B24**, 63–76.

Swanson, D. K., and R. C. Peterson (1980) Polyhedral volume calculations. *Canadian Mineralogist* **18**, 153–156.

Winter, J. K., S. Ghose and F. P. Okamura (1977) A high-temperature study of the thermal expansion and the anisotropy of the sodium atom in low albite. *Am. Mineral.* **62**, 921–931.

Willis, B. T. M., and A. W. Pryor (1975) *Thermal Vibrations in Crystallography*, Cambridge University Press, Cambridge, 280 pp.

Suppliers

The following list of suppliers is intended as a convenience for the reader. This is not a complete list of all firms that supply materials for high-temperature and high-pressure research. Inclusion in the list does not constitute an endorsement by the authors or Wiley International. All suppliers are United States firms unless otherwise noted.

Company Name and Address	*Materials Supplied*
Blake Industries, Inc. 660 Jerusalem Road Scotch Plains, New Jersey 07076	Single-crystal x-ray equipment, including diamond pressure cells and heaters
Charles Supper, Inc. Tech Circle Natick, Massachusetts 01760	X-ray cameras and accessories
Dubble Dee Diamond Corporation 2 West 46th Street New York, New York 10036	Diamond anvils
E. I. Dupont de Nemours and Company Wilmington, Delaware 19898	Kapton high-temperature plastic sheet
Enraf Nonius P.O. Box 438 2600AL Delft, The Netherlands	Single-crystal x-ray equipment, including diamond pressure cells and accessories
Fortafix Ltd Fengate Peterborough PE1 5BJ England	High-temperature cements

General Rand Corporation P.O. Box 6515 Bridgewater, New Jersey 08807	Silica glass capillaries
High-Pressure Diamond Optics 2741 East Calle los Altos Tuscon, Arizona 85718	Diamond-anvil pressure cells
International Nickel Corporation 1000 16th Street, N.W. Washington, DC 20036	Inconel metal for diamond-anvil cell gaskets
Kawecki Berylco Industries, Inc. P.O. Box 429 Hazelton, Pennsylvania 18201	Beryllium components
Lazare Kaplan and Sons Sample output from STRAIN: 107	Diamond anvils
Norton Company Industrial Ceramics Division Worcester, Massachusetts 01606	'Norbide' boron carbide
Sauereisen Cements Company 3389 Shapsburg Station Pittsburgh, Pennsylvania 15215	High-temperature furnace cements
Speedring Manufacturing Division 1635 Second Avenue, N.W. Cullman, Alabama 35055	Beryllium components
Zirconium Corporation of America P.O. Box 39217 Solon, Ohio 44139	High-temperature cements

A Program to Calculate the Strain Tensor from Two Sets of Unit-cell Parameters

by **Y. Ohashi***

```
C    ****************************************************************
C    STRAIN TENSOR CALCULATION PROGRAM FROM TWO SETS OF CELL PARAMETERS
C    VERSION 1 (1972)
C
C    CODED BY Y.OHASHI, GEOPHYSICAL LABORATORY, CARNEGIE INSTITUTION
C    OF WASHINGTON
C
C    ****************************************************************
C    INPUT DATA CARD FORMATS
C    ****************************************************************
C
C    1.TITLE CARD (18A4)
C
C       ANY HOLLERITH INFORMATION IN COLUMNS 1-72
C
C
C    2.CELL CARDS (3F8.0,3(F7.0,F5.0),F10.0,I1)
C
C       FIRST CARD FOR CELL PARAMETERS BEFORE DEFORMATION AND SECOND
C       CARD FOR THOSE AFTER DEFORMATION.
C
C       COLS.1-8,9-16,17-24 = A,B AND C IN ANGSTROMS
C       COLS.25-31,37-43,49-55 = ALPHA,BETA AND GAMMA IN DEGREES
C
C       IF ANGLES ARE IN DEGREES AND MINUTES,IN ADDITION THE FOLLOWING
C       COLUMNS CAN BE USED FOR FRACTIONAL PARTS.
C       COLS.32-36,44-48,56-60 = FRACTIONAL PARTS OF ALPHA,BETA AND
C                                GAMMA IN MINUTES
C       COLS.61-70 = TEMPERATURE (FOR THERMAL EXPANSION) OR COMPOSITION
C                    (FOR SOLID SOLUTION)  DIFFERENCE OF THIS QUANTITY
C                    BETWEEN FIRST AND SECOND CELL CARDS IS USED TO
C                    COMPUTE UNIT STRAIN
```

*Current Address: Department of Geology, University of Pennsylvania, Philadelphia, Pennsylvania, 19174.

```
C        COL.71  ICV(I) = 0  NO ERROR CARD FOR THIS CELL
C                        1  READ STANDARD DEVIATIONS OF CELL PARAMETERS
C                        2  READ VARIANCE-COVARIANCE MATRIX OF  CELL
C                           PARAMETERS
C
C     3.ERROR CARDS FOR EACH CELL
C     NEEDED ONLY IF ICV(I) IS NON-ZERO
C
C        IF ICV(I)=1, STANDARD DEVIATION CARD
C        IF ICV(I)=2, VARIANCE-COVARIANCE MATRIX CARDS
C
C     3-A STANDARD DEVIATION CARD (3F8.0,3(F7.0,F5.0))
C     NEEDED ONLY IF ICV(I)=1
C
C        COLS.1-8,9-16,17-24 = STANDARD DEVIATIONS OF A,B AND C IN
C                              ANGSTROMS
C        COLS.25-31,37-43,49-55 = STANDARD DEVIATIONS OF ALPHA,BETA AND
C                                 GAMMA IN DEGREES
C
C        IF STANDARD DEVIATIONS OF ANGLES ARE IN MINUTES, USE THE
C        FOLLOWING COLUMNS
C        COLS.32-36,44-48,56-60 = STANDARD DEVIATIONS OF ALPHA ,BETA AND
C                                 GAMMA IN MINUTES
C
C     3-B VARIANCE-COVARIANCE MATRIX CARDS (7E10.4)
C     NEEDED ONLY IF ICV(I)=2
C
C     THREE CARDS ARE REQUIRED TO INPUT 21 ELEMENTS OF THE TRIANGULAR
C     VARIANCE-COVARIANCE MATRIX FOR EACH CELL. UNITS OF ELEMENTS ARE
C     ANGSTROMS**2,RADIANS**2 OR ANGSTROMS*RADIANS.
C
C     CARD 1 (7E10.4)
C        A-A,A-B,A-C,A-ALPHA,A-BETA,A-GAMMA,AND B-B
C     CARD 2 (7E10.4)
C        B-C,B-ALPHA,B-BETA,B-GAMMA,C-C,C-ALPHA,AND C-BETA
C     CARD 3 (7E10.4)
C        C-GAMMA,ALPHA-ALPHA,ALPHA-BETA,ALPHA-GAMMA,BETA-BETA,BETA-GAMMA
C        AND GAMMA-GAMMA.
C
C     IF DESIRED, ANOTHER SET OF DATA CAN FOLLOW.
C
C
C     *****************************************************************
C     EXPLANATION OF PRINTOUT
C     *****************************************************************
C
C     1. ABC AND XYZ COORDINATE SYSTEMS
C        DIRECTION OF PRINCIPAL AXES OF STRAIN ELLIPSOID IS GIVEN IN
C        TERMS OF TWO COORDINATE SYSTEMS. A,B AND C ARE THE CRYSTALLO-
C        GRAPHIC AXES AND CAN BE OBLIQUE. X,Y AND Z ARE ORTHOGONAL AXES
C        SET UP IN THE PROGRAM. ANGULAR RELATIONS BETWEEN ABC AND XYZ
C        SYSTEMS ARE PRINTED.
C
C     2.DIMENSIONS OF STRAIN AND UNIT STRAIN
C        STRAIN IS DIMENSIONLESS (CHANGE IN LENGTH/ORIGINAL LENGTH), AND
C        UNIT STRAIN IS (STRAIN PER UNIT CHANGE OF TEMPERATURE OR COM-
C        POSITION)
C
C     3.POLAR COORDINATES AND STEREOGRAPHIC PROJECTION COORDINATES
C        SYSTEM XYZ IS USED FOR STEREO PROJECTION AXES WITH Z BEEING
C        THE AZIMUTHAL AXIS.
C        'RHO' IS AN AZIMUTHAL ANGLE AND 'PHI' IS A LONGITUDINAL ANGLE,
C        THUS X(RHO=90,PHI=0),Y(RHO=90,PHI=90) AND Z(RHO=0).
```

```
C
C
C       'PROJ.X' AND 'PROJ.Y' ARE PROJECTED COMPONENTS ON A STEREO-
C       GRAPHIC PROJECTION PLANE AND SCALED AS RADIUS=100.
C       '+' AND '-' DENOTE UPPER AND LOWER HEMISPHERES RESPECTIVELY.
C
C       TO OBTAIN STEREOGRAPHIC PROJECTION OF PRINCIPAL AXES AND ALSO
C       OF CRYSTALLOGRAPHIC AXES, PLOT 'PROJ.X' AND 'PROJ.Y' ON GRAPH
C       PAPER AND DRAW A CIRCLE WITH RADIUS 100.
C       STRAIN
        COMMON C,C1,QI,Q,ICV,VC,FO,DC,S
        DIMENSION TITLE(18),C(6),C1(6),Q(3,3),QI(3,3),E(3,3),DC(3,3)
        DIMENSION EVL(3),EVC(3,3),VC(2,6,6)
        DIMENSION AN(3),IAX(6),ICV(2),FO(13),S(13)
        DIMENSION ARRAY(6)
        DATA RAD/57.29579/
        DATA IN,IOUT/5,6/
        DATA IS1,IS2/2H +,2H -/,IAX/2H+X,2H+Y,2H+Z,2H+A,2H+B,2H+C/
        TAN(X)=SIN(X)/COS(X)
   10 FORMAT (18A4)
   20 FORMAT (1H1,18A4)
   30 FORMAT (3F8.0,3(F7.0,F5.0),F10.0,I1)
   40 FORMAT (1H0,'CELL PARAMETERS'/10X,'BEFORE',6(3X,F9.5),3X,'AT',
      1 3X,F7.2/10X,'AFTER',1X,6(3X,F9.5),3X,'AT',3X,F7.2)
   45 FORMAT (1H0,'STRAIN ELLIPSOID'/19X,'STRAIN',5X,'UNIT STRAIN',9X,
      1 'ANGLE WITH',14X,'RHO',7X,'PHI',5X,'PROJ.X',3X,'PROJ.Y',3X,
      2 'HEMISPHERE'/48X,'+X',7X,'+Y',7X,'+Z'/)
   50 FORMAT (1H ,9X,'AXIS',I2,1X,2(E11.4,2X),3F9.2,F8.2,3X,F7.2,2F9.1,
      1 7X,A2)
   52 FORMAT (1H ,9X,'AXIS',I2,1X,2(E11.4,'(',E10.3,')',1X),3(F9.2,
      1 '(',F6.2,')',2X))
   55 FORMAT (///,20X,'STRAIN (ERROR)',7X,'UNIT STRAIN (ERROR)',22X,
      1'ANGLE WITH' /70X,'+A   (ERROR)',8X,'+B   (ERROR)',8X,'+C  (ERROR)')
   60 FORMAT (1H0,'STRAIN TENSOR BASED ON XYZ'/3(10X,3E15.7/))
   65 FORMAT (1H0,'ANGLES BETWEEN XYZ AND ABC SYSTEMS'/20X,'+A',7X,'+B'
      1,7X,'+C'/)
   67 FORMAT (1H ,12X,A2,1X,3F9.2)
   70 FORMAT(1H ,9X,A2,' AXIS',26X,3F9.2,F8.2,3X,F7.2,2F9.1,7X,A2)
   71 FORMAT (1H ,9X,A2,' AXIS',26X,2(4X,'90.00'),1X,2(5X,'.00'),7X,'.00
      1',2(7X,'.0'),7X,' +')
   75 FORMAT(1H0,'CRYSTALLOGRAPHIC AXES')
   80 FORMAT (1H0,9X,'VOLUME',1X,2(E11.4,'(',E10.3,')',1X))
   85 FORMAT(3F8.0,3(F7.0,F5.0))
   86 FORMAT (7E10.4)
   90 FORMAT (1H0,'STANDARD DEVIATIONS FOR CELL',1X,I1/16X,6(3X,F9.5))
   91 FORMAT (1H0,'VARIANCE-COVARIANCE MATRIX FOR CELL ',I1/13X,'A(ANG.)
      1',9X,'B(ANG.)',9X,'C(ANG)',7X,'ALPHA(RAD.)',6X,'BETA(RAD.)',5X,'GA
      2MMA(RAD.)'/5X,'A',6(4X,E12.5)/5X,'B',20X,5(E12.5,4X)/5X,'C',36X,
      3 4(E12.5,4X)/5X,'ALPHA',48X,3(E12.5,4X)/5X,'BETA',65X,2(E12.5,4X)/
      4 5X,'GAMMA',80X,E12.5)
  100 READ(IN,10,END=500) TITLE
        WRITE (6,20) TITLE
        READ(IN,30) (C(I),I=1,3),ARRAY,T,ICV(1)
        IF (C(1).EQ.0.0) GO TO 500
        DO 101 I=1,3
  101 C(I+3)=ARRAY(2*I-1)+ARRAY(2*I)/60
        READ(IN,30) (C1(I),I=1,3),ARRAY,T,ICV(1)
        DO 1101 I=1,3
 1101 C1(I+3)=ARRAY(2*I-1)+ARRAY(2*I)/60.
        WRITE (6,40) C,T,C1,T1
        DO 108 I=1,2
        IF(ICV(I)-1) 108, 102, 104
  102 READ(IN,85) (VC(I,J,J),J=1,3),ARRAY
        DO 1103 J=1,3
 1103 VC(I,J+3,J+3)=ARRAY(2*J-1)+ARRAY(2*J)/60.
```

```
      WRITE (6,90) I,(VC(I,J,J),J=1,6)
      DO 103 J=1,3
      VC(I,J+3,J+3)=(VC(I,J+3,J+3)/RAD)**2
      VC(I,J,J)=VC(I,J,J)**2
  103 CONTINUE
      GO TO 108
  104 READ(IN,86) ((VC(I,J,K),K=J,6),J=1,6)
      WRITE (6,91) I,((VC(I,J,K),K=J,6),J=1,6)
      DO 109 J=1,5
      DO 109 K=J+1,6
      VC(I,K,J)=VC(I,J,K)
  109 CONTINUE
  108 CONTINUE
      DTEM=T1-T
      DO 120 I=1,3
      DO 110 J=1,3
      Q(I,J)=0.0
      QI(I,J)=0.0
      DC(I,J)=0.0
  110 CONTINUE
      C(I+3)=COS(C(I+3)/RAD)
      C1(I+3)=COS(C1(I+3)/RAD)
  120 CONTINUE
      TEMP=C(4)**2+C(5)**2+C(6)**2
      IOB=0
      IF (TEMP.GT.0.0001) IOB=1
      P=SQRT(1.-TEMP+2.*C(4)*C(5)*C(6))
      SA=SQRT(1.-C(4)**2)
      QI(1,1)=SA/(C(1)*P)
      QI(1,2)=(C(4)*C(5)-C(6))/(C(2)*P*SA)
      QI(1,3)=(C(6)*C(4)-C(5))/(C(3)*P*SA)
      QI(2,2)=1./(C(2)*SA)
      QI(2,3)=-C(4)/(C(3)*SA)
      QI(3,3)=1./C(3)
      IF (IOB.EQ.0) GO TO 130
      DC(1,1)=P/SA
      DC(2,1)=(C(6)-C(4)*C(5))/SA
      DC(3,1)=C(5)
      DC(2,2)=SA
      DC(3,2)=C(4)
      DC(3,3)=1.
      WRITE (6,65)
      DO 125 I=1,3
      DO 123 J=1,3
      AN(J)=RAD*ACOS(DC(I,J))
  123 CONTINUE
      WRITE (6,67) IAX(I),AN
  125 CONTINUE
  130 TEMP=C1(4)**2+C1(5)**2+C1(6)**2
      P=SQRT(1.-TEMP+2.*C1(4)*C1(5)*C1(6))
      SA=SQRT(1.-C1(4)**2)
      Q(1,1)=C1(1)*P/SA
      Q(1,2)=C1(1)*(C1(6)-C1(4)*C1(5))/SA
      Q(1,3)=C1(1)*C1(5)
      Q(2,2)=C1(2)*SA
      Q(2,3)=C1(2)*C1(4)
      Q(3,3)=C1(3)
C     QI*Q
      DO 150 I=1,3
      DO 150 J=1,3
      TEMP=0.0
      DO 140 K=1,3
      TEMP=TEMP+QI(I,K)*Q(K,J)
  140 CONTINUE
```

```
           E(I,J)=TEMP
   150 CONTINUE
C      SYMMETRIC TENSOR
           DO 160 I=1,3
           E(I,I)=E(I,I)-1.
   160 CONTINUE
           DO 170 I=1,2
           JJ=I+1
           DO 170 J=JJ,3
           E(I,J)=.5*(E(I,J)+E(J,I))
           E(J,I)=E(I,J)
   170 CONTINUE
           WRITE (6,60) ((E(I,J),J=1,3),I=1,3)
           CALL EIGEN1 (E,EVL,EVC)
           WRITE (6,45)
           DO 200 I=1,3
           DO 190 J=1,3
           AN(J)=RAD*ACOS(EVC(I,J))
   190 CONTINUE
C      RHO-PHI AND PROJECTION COORDINATES
           RHO=AN(3)
           PHI=ATAN2(EVC(I,2),EVC(I,1))
           IF(EVC(I,3)) 192,194,194
   192 IZ=IS2
           TEMP=TAN((180.-RHO)/(2.*RAD))*100
           GO TO 196
   194 IZ=IS1
           TEMP=TAN(RHO/(2.*RAD))*100
   196 PX=TEMP*COS(PHI)
           PY=TEMP*SIN(PHI)
           PHI=PHI*RAD
           EX=EVL(I)/DTEM
           WRITE (6,50) I,EVL(I),EX,AN,RHO,PHI,PX,PY,IZ
   200 CONTINUE
C      CALCULATE DIRECTIONS OF A,B,C AXES WITH RESPECT TO X,Y,Z AXES
           WRITE(6,75)
           DO 300 K=1,2
           DO 260 J=1,3
   260 AN(J)=ACOS(DC(J,K))*RAD
           RHO=AN(3)
           PHI=ATAN2(DC(2,K),DC(1,K))
           IF(DC(3,K)) 270,280,280
   270 IZ=IS2
           TEMP=TAN((180.-RHO)/(2.*RAD))*100.
           GO TO 290
   280 IZ=IS1
           TEMP=TAN(RHO/(2.*RAD))*100.
   290 PX=TEMP*COS(PHI)
           PY=TEMP*SIN(PHI)
           PHI=PHI*RAD
           WRITE(6,70) IAX(3+K),AN,RHO,PHI,PX,PY,IZ
   300 CONTINUE
           WRITE (6,71) IAX(6)
           WRITE (6,55)
           DO 220 I=1,3
           DO 215 J=1,3
           TEMP=0.0
           DO 210 K=1,3
           TEMP=TEMP+EVC(I,K)*DC(K,J)
   210 CONTINUE
           FO(3*J+I)=ACOS(TEMP)
   215 CONTINUE
           FO(I)=EVL(I)
```

```
  220 CONTINUE
      TEMP=EVL(1)+EVL(2)+EVL(3)
      FO(13)=TEMP
      ITEMP=ICV(1)+ICV(2)
      IF (ITEMP.NE.0) CALL ERROR1
      DO 250 I=1,3
      DO 240 J=1,3
      AN(J)=RAD*FO(3*J+I)
  240 CONTINUE
      TEMP=EVL(I)/DTEM
      TEMP1=S(I)/DTEM
      WRITE (6,52) I,EVL(I),S(I),TEMP,TEMP1,(AN(J),S(3*J+I),J=1,3)
  250 CONTINUE
      TEMP=FO(13)/DTEM
      TEMP1=S(13)/DTEM
      WRITE (6,80) FO(13),S(13),TEMP,TEMP1
      GO TO 100
  500 STOP
      END
      SUBROUTINE ERROR1
      COMMON     C,QI,Q,ICV,VC,FO,DC,S
      DIMENSION FO(13),F(13),D(12,13),C(12),A(3,3),B(3,3),XX(3),VV(3,3),
     1 QI(3,3),Q(3,3),DC(3,3),R(3,3),VC(2,6,6),E(3,3),S(13),ICV(2)
      EQUIVALENCE (F(1),XX(1)),(F(4),VV(1,1))
      DATA RAD/57.29578/,CONST/.01/
C     DERIVATIVES OF EIGEN VALUES AND DIRECTION COSINES WITH RESPECT TO
C     CELL CONSTANTS
      DO 200 I=1,2
      L=6*I-3
      TEMP=C(L+1)**2+C(L+2)**2+C(L+3)**2
      P=SQRT(1.-TEMP+2.*C(L+1)*C(L+2)*C(L+3))
      SA=SQRT(1.-C(L+1)**2)
      IF (I.EQ.2) GO TO 102
      DO 100 L=1,3
      DO 100 N=L,3
      B(L,N)=Q(L,N)
  100 CONTINUE
  102 DO 104 L=1,3
      DO 104 N=L,3
C     QI,Q=UPPER TRIANGULAR MATRICES , DC=LOWER TRIANGULAR MATRIX
      A(L,N)=QI(L,N)
      R(N,L)=DC(N,L)
  104 CONTINUE
      DO 200 J=1,6
      IJ=(I-1)*6+J
      IF (VC(I,J,J)-1.E-10) 105,105,107
  105 DO 106 K=1,13
      D(IJ,K)=0.0
  106 CONTINUE
      GO TO 200
  107 DEL=CONST*SQRT(VC(I,J,J))
      IF (J-3) 110,110,120
  110 C(IJ)=C(IJ)+DEL
      GO TO 130
  120 C(IJ)=COS(ACOS(C(IJ))+DEL)
      L=6*I-3
      TEMP=C(L+1)**2+C(L+2)**2+C(L+3)**2
      P=SQRT(1.-TEMP+2.*C(L+1)*C(L+2)*C(L+3))
      SA=SQRT(1.-C(L+1)**2)
C     CHANGE OF QI OR Q MATRICES
  130 GO TO (140,150),I
  140 DO 145 L=1,3
      DO 145 N=L,3
```

```
      A(L,N)=QI(L,N)
      R(N,L)=DC(N,L)
145 CONTINUE
      GO TO (11,12,13,14,15,15),J
 11 A(1,1)=SA/(C(1)*P)
      GO TO 160
 12 A(1,2)=(C(4)*C(5)-C(6))/(C(2)*P*SA)
      A(2,2)=1./(C(2)*SA)
      GO TO 160
 13 A(1,3)=(C(6)*C(4)-C(5))/(C(3)*P*SA)
      A(2,3)=-C(4)/(C(3)*SA)
      A(3,3)=1./C(3)
      GO TO 160
 14 A(2,2)=1./(C(2)*SA)
      A(2,3)=-C(4)/(C(3)*SA)
      R(3,2)=C(4)
 15 A(1,1)=SA/(C(1)*P)
      A(1,2)=(C(4)*C(5)-C(6))/(C(2)*P*SA)
      A(1,3)=(C(6)*C(4)-C(5))/(C(3)*P*SA)
      R(1,1)=P/SA
      R(2,1)=(C(6)-C(4)*C(5))/SA
      R(3,1)=C(5)
      R(2,2)=SA
      GO TO 160
150 DO 155 L=1,3
      DO 155 N=L,3
      B(L,N)=Q(L,N)
155 CONTINUE
      GO TO (21,22,23,24,25,26),J
 21 B(1,1)=C(7)*P/SA
      B(1,2)=C(7)*(C(12)-C(10)*C(11))/SA
      B(1,3)=C(7)*C(11)
      GO TO 160
 22 B(2,2)=C(8)*SA
      B(2,3)=C(8)*C(10)
      GO TO 160
 23 B(3,3)=C(9)
      GO TO 160
 24 B(1,1)=C(7)*P/SA
      B(1,2)=C(7)*(C(12)-C(10)*C(11))/SA
      B(2,2)=C(8)*SA
      B(2,3)=C(8)*C(10)
      GO TO 160
 25 B(1,3)=C(7)*C(11)
 26 B(1,1)=C(7)*P/SA
      B(1,2)=C(7)*(C(12)-C(10)*C(11))/SA
160 DO 170 L=1,3
      DO 170 N=L,3
      TEMP=0.0
      DO 165 M=L,N
      TEMP=TEMP+A(L,M)*B(M,N)
165 CONTINUE
      E(L,N)=TEMP
170 CONTINUE
      DO 175 L=1,3
      E(L,L)=E(L,L)-1.
175 CONTINUE
      DO 180 L=1,2
      DO 180 N=L+1,3
      E(L,N)=.5*E(L,N)
      E(N,L)=E(L,N)
180 CONTINUE
```

```
      CALL EIGEN1(E,XX,VV)
      DO 184 L=1,3
      DO 184 N=1,3
      TEMP=0.0
      DO 182 K=N,3
      TEMP=TEMP+VV(L,K)*R(K,N)
  182 CONTINUE
      F(3*N+L)=ACOS(TEMP)
  184 CONTINUE
      F(13)=F(1)+F(2)+F(3)
      DO 185 K=1,13
      D(IJ,K)=(F(K)-FO(K))/DEL
  185 CONTINUE
      IF (J-3) 190,190,195
  190 C(IJ)=C(IJ)-DEL
      GO TO 200
  195 C(IJ)=COS(ACOS(C(IJ))-DEL)
  200 CONTINUE
      DO 205 K=1,13
      S(K)=0.0
  205 CONTINUE
      DO 300 I=1,2
C     VARIANCE CALCULATION
      L=6*(I-1)
      IF (ICV(I)-1) 300,208,250
C     ZERO COVARIANCE CASE
  208 DO 220 K=1,13
      TEMP=0.0
      DO 210 IP=1,6
      LP=L+IP
      TEMP=TEMP+D(LP,K)**2*VC(I,IP,IP)
  210 CONTINUE
      S(K)=S(K)+TEMP
  220 CONTINUE
      GO TO 300
C     NON-ZERO COVARIANCE CASE
  250 DO 270 K=1,13
      TEMP=0.0
      DO 260 IP=1,6
      DO 260 IQ=1,6
      LP=L+IP
      LQ=L+IQ
      TEMP=TEMP+D(LP,K)*D(LQ,K)*VC(I,IP,IQ)
  260 CONTINUE
      S(K)=TEMP+S(K)
  270 CONTINUE
  300 CONTINUE
      DO 310 K=1,3
      S(K)=SQRT(S(K))
  310 CONTINUE
      DO 320 K=4,12
      S(K)=RAD*SQRT(S(K))
  320 CONTINUE
      S(13)=SQRT(S(13))
      RETURN
      END
      SUBROUTINE EIGEN1 (F,X,V)
C     DIAGONALIZATION OF MATRIX F
      DIMENSION X(3),AM(3,3),F(3,3),V(3,3)
      DATA PI/3.141592/
   10 FORMAT (1H0,'NON-REAL SOLUTIONS')
      F2=-F(1,1)-F(2,2)-F(3,3)
```

```
      F1=F(1,1)*F(2,2)+F(2,2)*F(3,3)+F(3,3)*F(1,1)
     1 -F(2,3)*F(3,2)-F(1,2)*F(2,1)-F(1,3)*F(3,1)
      F0=F(1,1)*F(2,3)*F(3,2)-F(1,1)*F(2,2)*F(3,3)
     1 +F(1,2)*F(2,1)*F(3,3)-F(1,2)*F(2,3)*F(3,1)
     2 +F(1,3)*F(2,2)*F(3,1)-F(1,3)*F(2,1)*F(3,2)
      P=-F2**2/3.+F1
      Q=2*(F2/3)**3-F2*F1/3+F0
      Q1=(P/3)**3+(Q/2)**2
      IF (Q1) 150,140,140
  140 WRITE (6,10)
      CALL EXIT
  150 P1=SQRT(-P/3)
      TEMP=ACOS(-Q/(2*P1**3))
      X(1)=-F2/3+2*P1*COS(TEMP/3)
      X(2)=-F2/3-2*P1*COS((TEMP+PI)/3)
      X(3)=-F2/3-2*P1*COS((TEMP-PI)/3)
      DO 280 I=1,2
      JJ=I+1
      DO 270 J=JJ,3
      IF (X(I).GT.X(J)) GO TO 270
      X1=X(I)
      X(I)=X(J)
      X(J)=X1
  270 CONTINUE
  280 CONTINUE
      DO 470 J=1,3
      DO 306 IH=1,3
      DO 306 K=1,3
      AM(IH,K)=F(IH,K)
  306 CONTINUE
      DO 330 I=1,3
      AM(I,I)=AM(I,I)-X(J)
  330 CONTINUE
      DET=0.0
      DO 380 IH=1,2
      KK=IH+1
      DO 380 K=KK,3
      D1=AM(IH,IH)*AM(K,K)-AM(IH,K)*AM(K,IH)
      IF (ABS(D1)-ABS(DET)) 380,350,350
  350 DET=D1
      IH1=IH
      K1=K
  380 CONTINUE
      I1=IH1
      I2=K1
      I3=6-IH1-K1
      V(J,I3)=1
      V(J,I2)=(AM(I1,I3)*AM(I2,I1)-AM(I1,I1)*AM(I2,I3))/DET
      V(J,I1)=(AM(I1,I2)*AM(I2,I3)-AM(I1,I3)*AM(I2,I2))/DET
      TEMP=V(J,1)**2+V(J,2)**2+V(J,3)**2
      TEMP=SQRT(TEMP)
      DO 475 K=1,3
      V(J,K)=V(J,K)/TEMP
  475 CONTINUE
C     EIGEN VECTORS V(I,J) ROW-WISE
  470 CONTINUE
      RETURN
      END
```

Sample Strain Calculation

Sample input and output for the calculation of a strain ellipsoid of unit-cell deformation are listed below. Unit-cell dimensions of $NaAlSi_3O_8$, low albite, at 26°C and 1127°C are taken from Stewart and von Limbach (1967). Note that whereas all three unit-cell axes expand with temperature in this triclinic mineral, the minor axis of the thermal expansion ellipsoid is actually negative.

REFERENCE TO APPENDIX II

Stewart, D. B., and D. von Limbach (1967) Thermal expansion of low and high albite. *Am. Mineral.* **52**, 389–413.

Sample input for program STRAIN:

```
LOW ALBITE UNIT-CELL STRAIN.  26 TO 1127°C (STEWART + VON LIMBACH, 1967)
 8.141   12.790   7.157   94.225      116.597      87.772      26.0       1
 8.278   12.863   7.180   92.780      116.045      87.718      1127.0     1
 0.007    0.009   0.004    0.058        0.048       0.052
 0.006    0.009   0.003    0.067        0.037       0.055
```

Sample output from STRAIN:

```
LOW ALBITE UNIT-CELL STRAIN.  26 TO 1127 C (STEWART + VON LIMBACH, 1967)

CELL PARAMETERS
   BEFORE   8.14100   12.79000   7.15700   94.22500   116.59700   87.77200   AT     26.00
   AFTER    8.27800   12.86300   7.18000   92.78000   116.04500   87.71800   AT   1127.00

STANDARD DEVIATIONS FOR CELL 1
            0.00700    0.00900   0.00400    0.05800    0.04800    0.05200

STANDARD DEVIATIONS FOR CELL 2
            0.00600    0.00900   0.00300    0.06700    0.03700    0.05500

ANGLES BETWEEN XYZ AND ABC SYSTEMS
          +A       +B       +C
   +X    26.60    90.00    90.00
   +Y    89.66     4.22    90.00
   +Z   116.60    94.22     0.00

STRAIN TENSOR BASED ON XYZ
   0.2149069E-01   0.7216586E-02   0.1418307E-02
   0.7216586E-02   0.7261276E-02   0.1260034E-01
   0.1418307E-02   0.1260034E-01   0.3213644E-02

STRAIN ELLIPSOID   STRAIN       UNIT STRAIN    ANGLE WITH               RHO      PHI     PROJ.X  PROJ.Y  HEMISPHERE
                                              +X      +Y      +Z
   AXIS 1    0.2663E-01   0.2419E-04    36.88   59.54   71.27   71.27    32.36    60.5     38.4      +
   AXIS 2    0.1332E-01   0.1210E-04   125.89   56.90   53.24   53.24   137.04   -36.7     34.2      +
   AXIS 3   -0.7989E-02  -0.7256E-05    82.65  131.84   42.78   42.78   -79.14     7.4    -38.5      +

CRYSTALLOGRAPHIC AXES
                                              +X      +Y      +Z
   +A AXIS                                   26.60   89.66  116.60   116.60     0.38    61.8      0.4      -
   +B AXIS                                   90.00    4.22   94.22    94.22    90.00     0.0     92.9      +
   +C AXIS                                   90.00   90.00     .00      .00      .00      .0       .0      +

                  STRAIN (ERROR)          UNIT STRAIN (ERROR)            ANGLE WITH
                                                                     +A (ERROR)      +B (ERROR)      +C (ERROR)
   AXIS 1    0.2663E-01( 0.125E-01)    0.2419E-04( 0.113E-05)    54.94( 3.49)    61.19( 2.44)    71.27( 2.31)
   AXIS 2    0.1332E-01( 0.900E-03)    0.1210E-04( 0.818E-06)   142.09( 3.56)    59.97( 2.63)    53.24( 1.96)
   AXIS 3   -0.7989E-02( 0.101E-02)   -0.7256E-05( 0.915E-06)   102.60( 1.56)   135.99( 1.40)    42.78( 1.53)

   VOLUME   0.3197E-01( 0.176E-02)    0.2903E-04( 0.160E-05)
```

A Program to Calculate Polyhedral Volumes and Polyhedral Distortion Parameters from a Set of Atomic Coordinates and a Unit Cell

```
C VOLCAL.FTN
C
C PROGRAM TO CALCULATE POLYHEDRAL VOLUMES AND DISTORTION PARAMETERS
C     WITH ERRORS (IF DESIRED).
C
C ORIGINAL PROGRAM BY L. W. FINGER,  9/21/71
C MODIFIED 10/24/73 BY Y. OHASHI
C MODIFIED 6/20/79  BY L. W. FINGER - MADE INTERACTIVE AND ERRORS ADDED
C
C     ****************************************************************
C     INPUT DATA FORMAT IF READ FROM FILE
C     ****************************************************************
C
C     1. TITLE CARD (18A4)
C     2. CELL CARD (6F8.0,2I4)
C        COLS.1-8,9-16,17-24 = A,B AND C
C        COLS.25-32,33-40,41-48 = ALPHA,BETA AND GAMMA IN DEG.
C        COLS.49-52  MP  =  NUMBER OF POLYHEDRA TO BE PROCESSED
C       COLS.53-56  IER =  NON-ZERO IF ERRORS TO BE CALCULATED
C     3. CELL VARIANCE-COVARIANCE MATRIX (6F10.0) (NEEDED ONLY IF IER
C  NON-ZERO)
C        THIS INPUT IS SIX LINES WITH ONE ROW OF MATRIX/LINE.
C     4. CENTRAL ATOM CARD (3F9.5,I4,2A4)
C        COLS.1-9,10-18,19-27  =  X,Y AND Z
C        COLS.28-31  NP  =  NUMBER OF COORDINATING ATOMS
C        COLS.32-39  SITE NAME
C     5. CENTRAL ATOM SIGMA CARDS (3F9.5) (NEEDED ONLY IF IER NON-ZERO)
C     COLS.1-9,10-18,19-27 = SIGMA X, Y, AND Z
C     6. COORDINATING ATOM CARDS (3F9.5,2A4)
C        COLS.1-9,10-18,19-27  =  X,Y AND Z
C        COLS.28-35   SITE NAME
C     7. COORDINATING ATOM SIGMA CARDS (3F9.5) NEEDED ONLY IF IER
C  NON-ZERO)
C        COLS.1-9,10-18,19-27 = SIGMA X, Y, AND Z
C
C     CARDS 6 AND 7 SHOULD BE REPEATED FOR EACH COORDINATING ATOM
```

```
C
C       IF DESIRED, ANOTHER SET OF DATA CAN FOLLOW
C
C       ******************************************************************
C
        LOGICAL*1 IYES,INO,IRESP,IPAGE
        DIMENSION TITLE(18),A(6),VARA(6,6),SITE(2),XC(3),SXC(3)
        DIMENSION XP(3,20),SXP(3,20),DERDA(6,3),AX(3),SIGMA(3)
       1,X(3),PRM(3)
        DATA IYES,INO,IPAGE,IN/1HY,1HN,1H ,4/
C
90      TYPE 1
1       FORMAT('0POLYHEDRAL VOLUME AND DISTORTION PARAMETER PROGRAM'
       1/'$ARE DATA TO BE READ FROM FILE(Y,N)?')
        ACCEPT 8,IRESP
        IFILE=1
        IF(IRESP.EQ.IYES)GO TO 1000
        IF(IRESP.NE.INO)GO TO 90
        IFILE=0
        TYPE 2
2       FORMAT('$TYPE TITLE?')
        ACCEPT 4,TITLE
4       FORMAT(18A4)
        TYPE 6
6       FORMAT('$ARE ERRORS TO BE CALCULATED(Y,N)?')
100     ACCEPT 8,IRESP
8       FORMAT(A1)
        IER=1
        IF(IRESP.EQ.IYES)GO TO 110
        IF(IRESP.NE.INO)GO TO 100
        IER=0
110     TYPE 10
10      FORMAT('$TYPE CELL CONSTANTS?')
        ACCEPT 12,A
12      FORMAT(6F10.0)
        IF(IER.EQ.0)GO TO 120
        TYPE 14
14      FORMAT(' TYPE VARIANCE MATRIX FOR CELL DATA (1 ROW PER LINE)')
        ACCEPT 12,VARA
120     TYPE 16
16      FORMAT('$TYPE NO. OF COORDINATING IONS?')
        ACCEPT 18,NP
18      FORMAT(I6)
        IF(NP.EQ.0)CALL EXIT
        WRITE(6,5)IPAGE,TITLE
5       FORMAT(A1,18A4)
        IPAGE='1'
        WRITE(6,13)A
13      FORMAT('0CELL CONSTANTS',6F10.5)
        IF(IER.NE.0)WRITE(6,15)VARA
15      FORMAT('0CELL COVARIANCE MATRIX'/(6E13.6))
        TYPE 20
20      FORMAT('$TYPE CATION COORDS. AND SITE NAME?')
        ACCEPT 22,XC,SITE
22      FORMAT(3F9.0,2A4)
        WRITE(6,23)SITE,XC
23      FORMAT('0CATION: ',2A4,3F10.5)
        IF(IER.EQ.0)GO TO 130
        TYPE 24
24      FORMAT('$TYPE SIGMAS FOR SITE?')
        ACCEPT 22,SXC
        WRITE(6,25)SXC
25      FORMAT(5X,'SIGMAS',6X,3F10.5)
```

```
130    WRITE(6,27)
27     FORMAT('0ANIONS')
       DO 140 I=1,NP
       TYPE 26,I
26     FORMAT('$TYPE ANION COORDS. AND LABEL FOR NO.',I3,'?')
       ACCEPT 22,(XP(J,I),J=1,3),SITE
       WRITE(6,29)I,SITE,(XP(J,I),J=1,3)
29     FORMAT(3X,I5,1X,2A4,3F10.5)
       IF(IER.EQ.0)GO TO 140
       TYPE 28
28     FORMAT('$SIGMAS?')
       ACCEPT 22,(SXP(J,I),J=1,3)
       WRITE(6,25)(SXP(J,I),J=1,3)
140    CONTINUE
       GO TO 1410
C
C READ DATA FROM FILE
C
1000   TYPE 40
40     FORMAT('$TYPE INPUT FILE NAME?')
       ACCEPT 42,I,TITLE
42     FORMAT(Q,18A4)
       CALL ASSIGN(IN,TITLE,I)
1010   READ(IN,4,END=260)TITLE
       READ(IN,44)A,MP,IER
       NCOUNT=0
44     FORMAT(6F8.0,2I4)
       WRITE(6,5)IPAGE,TITLE
       IPAGE='1'
       WRITE(6,13)A
       IF(IER.EQ.0)GO TO 1020
       READ(IN,12)VARA
       WRITE(6,15)VARA
1020   READ(IN,46)XC,NP,SITE
46     FORMAT(3F9.0,I4,2A4)
       WRITE(6,23)SITE,XC
       IF(IER.EQ.0)GO TO 1030
       READ(IN,22)SXC
       WRITE(6,25)SXC
1030   DO 1040 I=1,NP
       READ(IN,22)(XP(J,I),J=1,3),SITE
       WRITE(6,29)I,SITE,(XP(J,I),J=1,3)
       IF(IER.EQ.0)GO TO 1040
       READ(IN,22)(SXP(J,I),J=1,3)
       WRITE(6,25)(SXP(J,I),J=1,3)
1040   CONTINUE
C
1410   CALL PRMCAL(0,A,A(4),XC,NP,XP,PRM(1),PRM(2),PRM(3))
       IF(IER.EQ.0)GO TO 250
C FUNCTIONS CALULATED - NOW GET ERRORS
       DO 142 I=1,3
142    SIGMA(I)=0.0
       DO 155 I=1,6
       DX=SQRT(VARA(I,I))*.1
       IF(DX.EQ.0.0)GO TO 155
       A(I)=A(I)+DX
       CALL PRMCAL(1,A,A(4),XC,NP,XP,X(1),X(2),X(3))
       A(I)=A(I)-DX
       DO 150 J=1,3
150    DERDA(I,J)=(X(J)-PRM(J))/DX
155    CONTINUE
       DO 170 I=1,6
```

```
          DO 170 J=1,3
          AX(I)=0.0
          DO 160 K=1,6
  160     AX(I)=AX(I)+DERDA(I,J)*VARA(K,I)
  170     SIGMA(J)=SIGMA(J)+AX(I)*DERDA(I,J)
          DO 225 I=1,3
          DX=XC(I)+.1*SXC(I)
          IF(DX.EQ.0.0)GO TO 225
          XC(I)=XC(I)+DX
          CALL PRMCAL(1,A,A(4),XC,NP,XP,X(1),X(2),X(3))
          XC(I)=XC(I)-DX
          DO 220 J=1,3
  220     SIGMA(J)=SIGMA(J)+((X(J)-PRM(J))/DX*SXC(I))**2
  225     CONTINUE
          DO 231 I=1,NP
          DO 231 J=1,3
          DX=XP(J,I)+0.1*SXP(J,I)
          IF(DX.EQ.0.0)GO TO 231
          XP(J,I)=XP(J,I)+DX
          CALL PRMCAL(1,A,A(4),XC,NP,XP,X(1),X(2),X(3))
          XP(J,I)=XP(J,I)-DX
          DO 230 K=1,3
  230     SIGMA(K)=SIGMA(K)+((X(K)-PRM(K))/DX*SXP(J,I))**2
  231     CONTINUE
          DO 232 I=1,3
  232     SIGMA(I)=SQRT(SIGMA(I))
          IF(NP.NE.4.AND.NP.NE.6)GO TO 240
          WRITE(6,30)SIGMA
   30     FORMAT(' SIGMA(S) ',3F12.5)
          GO TO 250
  240     WRITE(6,30)SIGMA(1)
  250     IF(IFILE.EQ.0)GO TO 120
          NCOUNT=NCOUNT+1
          IF(NCOUNT-MP)1020,1010,1010
  260     CALL EXIT
          END
          SUBROUTINE PRMCAL(IORG,A,ANG1,XC,NP,YP,VOL,QE,AV)
          DIMENSION XC(3),A(3),ANG1(3),ANG(3),XP(3,20),VXYZ(6),D(3)
         1,DM(20),IH(3)
          DIMENSION YP(3,20)
C
          DATA IOUT/6/
          DATA PI/3.141592/
C
    6 FORMAT(1H0,'POLYHEDRAL VOLUME',F13.5/5X,'QUAD. ELONG.',F14.5/5X,
       1'ANGLE VARIANCE',F12.5)
    8 FORMAT (1H0,'POLYHEDRAL VOLUME',F12.5)
   15 FORMAT(1H0,10X,'ATOMS',13X,'DISTANCES',20X,'ANGLE'/12X,'I    J',
       1 7X,'0-I',9X,'0-J',9X,'I-J',7X,'I-0-J'/)
   20 FORMAT(1H ,8X,2I4,3F12.5,F12.3)
   25 FORMAT(1H0,12X,'FACE DEFINED',10X,'PLANE NORMAL',5X,'AREA'/13X,
       1 'BY POINTS',7X,'OX',6X,'OY',6X,'OZ'/)
   26 FORMAT (1H ,10X,3I3,4X,3F8.5,4X,F10.5)
C
          DO 100 I=1,3
  100 ANG(I)=COS(ANG1(I)*PI/180.0)
          SINA=SQRT(1.0-ANG(1)**2)
          ASTAR=SINA/(A(1)*SQRT(1.0-ANG(1)**2-ANG(2)**2-ANG(3)**2+2.0*ANG(1)
         1*ANG(2)*ANG(3)))
C        CONVERT POINTS TO ORTHOGONAL COORDS.
          DO 130 I=1,NP
C        REFER TO ORIGIN AT CENTRAL ATOM
          DO 120 J=1,3
```

```
      120 XP(J,I)=YP(J,I)-XC(J)
          XP(3,I)=A(3)*XP(3,I)+XP(2,I)*A(2)*ANG(1)+XP(1,I)*A(1)*ANG(2)
          XP(2,I)=XP(2,I)*A(2)*SINA+XP(1,I)*A(1)*(ANG(3)-ANG(2)*ANG(1))/SINA
          XP(1,I)=XP(1,I)/ASTAR
          DM(I)=SQRT(XP(1,I)**2+XP(2,I)**2+XP(3,I)**2)
      130 CONTINUE
C         DISTANCES AND ANGLES
          IF(IORG.EQ.0)WRITE (IOUT,15)
          DO 145 I=1,NP-1
          DO 145 J=I+1,NP
          TEMPA=0.
          TEMPB=0.
          DO 135 K=1,3
          TEMPA=TEMPA+(XP(K,I)-XP(K,J))**2
      135 TEMPB=TEMPB+XP(K,I)*XP(K,J)
          DIS=SQRT(TEMPA)
          ANGL=TEMPB/(DM(I)*DM(J))
          SUMTH=1.0-ANGL*ANGL
          IF(SUMTH.LT.0.0)SUMTH=0.0
          ANGL=57.2958*ATAN2(SQRT(SUMTH),ANGL)
          IF(IORG.EQ.0)WRITE (IOUT,20) I,J,DM(I),DM(J),DIS,ANGL
      145 CONTINUE
          IF(IORG.EQ.0)WRITE(IOUT,25)
C             FIND FACES    - CHOOSE ANY THREE POINTS
          VOL=0.0
          NP1=NP-1
          NP2=NP-2
          SUMTH =0.0
          SUMTH2=0.0
          NA=0
          DO 180 I=1,NP2
          IH(1)=I
          I1=I+1
          DO 180 J=I1,NP1
          J1=J+1
          IH(2)=J
          VXYZ(1)=XP(1,J)-XP(1,I)
          VXYZ(2)=XP(2,J)-XP(2,I)
          VXYZ(3)=XP(3,J)-XP(3,I)
          DO 180 K=J1,NP
          IH(3)=K
          NS=0
          VXYZ(4)=XP(1,K)-XP(1,I)
          VXYZ(5)=XP(2,K)-XP(2,I)
          VXYZ(6)=XP(3,K)-XP(3,I)
          D(1)=VXYZ(2)*VXYZ(6)-VXYZ(5)*VXYZ(3)
          D(2)=VXYZ(4)*VXYZ(3)-VXYZ(1)*VXYZ(6)
          D(3)=VXYZ(1)*VXYZ(5)-VXYZ(4)*VXYZ(2)
          AREA=0.5*SQRT(D(1)**2+D(2)**2+D(3)**2)
          Z0=0.5*(XP(1,I)*D(1)+XP(2,I)*D(2)+XP(3,I)*D(3))/AREA
C             CHECK FOR AND AVOID PLANE THROUGH ORIGIN
          IF(ABS(Z0).LT.1.0E-5) GO TO 180
          NS=SIGN(1.05,Z0)
          DO 170 L=1,NP
          IF(L.EQ.I.OR.L.EQ.J.OR.L.EQ.K) GO TO 170
C             CALCULATE DISTANCE OF POINT L FROM PLANE OF IJK
          Z=0.5*((XP(1,I)-XP(1,L))*D(1)+(XP(2,I)-XP(2,L))*D(2)+(XP(3,I)-XP(3
         1,L))*D(3))/AREA
          IF(NS)150,160,160
C             FIRST Z NEGATIVE, CHECK THAT THIS ONE IS ALSO
      150 IF(Z-1.0E-5) 170,170,180
C             HERE IF FIRST Z POSITIVE
      160 IF(Z+1.0E-5)180,170,170
```

```
  170 CONTINUE
C         ALL POINTS ON SAME SIDE,   THUS IJK ARE FACE
C
C     DIRECTION COSINES OF PLANE NORMAL
      D(1)=D(1)/(AREA*2.)
      D(2)=D(2)/(AREA*2.)
      D(3)=D(3)/(AREA*2.)
      IF(IORG.EQ.0)WRITE (IOUT,26) I,J,K,D,AREA
      VOL=VOL+AREA*ABS(Z0)/3.0
      DO 172 L=1,2
      L1=L+1
      NM=IH(L)
      DO 172 M=L1,3
      MN=IH(M)
      TEMP=0.0
      DO 171 N=1,3
  171 TEMP=TEMP+XP(N,NM)*XP(N,MN)
      TEMP=TEMP/(DM(NM)*DM(MN))
      TEMP=57.2958*ATAN2(SQRT(1.0-TEMP*TEMP),TEMP)
      SUMTH=SUMTH+TEMP
  172 SUMTH2=SUMTH2+TEMP**2
      NA=NA+3
  180 CONTINUE
      IF((NP-4)*(NP-6))230,190,230
C         ALL FACES FOUND, CALCULATE QUADRATIC ELONGATION FOR TETR AND OC
  190 SUM=0.0
      DO 200 I=1,NP
  200 SUM=SUM+DM(I)**2
      IF(NP.EQ.4) GO TO 210
C         SET UP FOR OCTAHEDRON
      CONS=0.75
      CONS2=90.0
      GO TO 220
C         SET UP FOR TETRAHEDRON
  210 CONS=9.0*SQRT(3.0)/8.0
      CONS2=109.045
  220 VLO=EXP(2.0*ALOG(CONS*VOL)/3.0)
      QE=SUM/(NP*VLO)
      NA=NA/2
      SUMTH2=0.5*SUMTH2
      SUMTH =0.5*SUMTH
      SIGA=(SUMTH2-2.0*CONS2*SUMTH+NA*CONS2**2)/(NA-1)
      IF(IORG.EQ.0)WRITE(IOUT,6) VOL,QE,SIGA
      GO TO 240
  230 IF(IORG.EQ.0)WRITE(IOUT,8) VOL
  240 CONTINUE
      RETURN
      END
```

Sample Polyhedral Volume Calculation

Sample input and output for the calculation of polyhedral volumes of cation sites in $ZrSiO_4$ (zircon) are listed below. The program, VOLCAL, calculates polyhedral volumes for all coordination groups, but polyhedral distortion indices are generated only for tetrahedral and octahedral cases. In zircon, therefore, volume only is reported for the 8-coordinated zircon site, whereas volume, quadratic elongation and bond angle variance appear for the silicon tetrahedron.

The error on bond angle variance is calculated as zero because unit-cell covariance terms are unknown (see text).

Data on $ZrSiO_4$ at 23 °C and 1 bar are from Hazen and Finger (1979).

REFERENCE TO APPENDIX III

Hazen, R. M., and L. W. Finger (1979) Crystal structure and compressibility of zircon at high pressure. *Am. Mineral.* **64**, 196–201.

Sample input for VOLCAL:

```
ZIRCON ZR AND SI POLYHEDRA AT 23C, 1 BAR.  (HAZEN + FINGER, 1979)
6.6042  6.6042  5.9796  90.0      90.0      90.0        2   1
0.000000160.00000016        0        0        0        0
0.000000160.00000016        0        0        0        0
        0          00.00000009        0        0        0
        0        0        0        0        0        0
        0        0        0        0        0        0
        0        0        0        0        0        0
0.0      0.75     0.125      8       ZR
0.0      0.0      0.0
0.0      1.0660   0.1951       O
0.0      0.0004   0.0004
0.0      0.4340   0.1951       O
0.0      0.0004   0.0004
0.0      0.9340  -.1951        O
0.0      0.0004   0.0004
0.0      0.5660  -.1951        O
0.0      0.0004   0.0004
-.1840   0.75     0.4451       O
0.0004   0.0      0.0004
0.1840   0.75     0.4451       O
0.0004   0.0      0.0004
-.3160   0.75     0.0549       O
0.0004   0.0      0.0004
0.3160   0.75     0.0549       O
0.0004   0.0      0.0004
0.0      0.25     0.375      4       SI
0.0      0.0      0.0
0.0      0.4340   0.1951       O
0.0      0.0004   0.0004
0.0      0.0660   0.1951       O
0.0      0.0004   0.0004
0.1840   0.25     0.5549       O
0.0004   0.0      0.0004
-.1840   0.25     0.5549       O
0.0004   0.0      0.0004
```

Sample output from VOLCAL:

```
ZIRCON ZR AND SI POLYHEDRA AT 23C, 1 BAR.  (HAZEN + FINGER, 1979)

CELL CONSTANTS    6.60420    6.60420    5.97960   90.00000   90.00000   90.00000

CELL COVARIANCE MATRIX
0.160000E-06 0.160000E-06 0.000000E+00 0.000000E+00 0.000000E+00 0.000000E+00
0.160000E-06 0.160000E-06 0.000000E+00 0.000000E+00 0.000000E+00 0.000000E+00
```

```
0.000000E+00 0.000000E+00 0.900000E-07 0.000000E+00 0.000000E+00 0.000000E+00
0.000000E+00 0.000000E+00 0.000000E+00 0.000000E+00 0.000000E+00 0.000000E+00
0.000000E+00 0.000000E+00 0.000000E+00 0.000000E+00 0.000000E+00 0.000000E+00
0.000000E+00 0.000000E+00 0.000000E+00 0.000000E+00 0.000000E+00 0.000000E+00
```

CATION:	ZR	0.00000	0.75000	0.12500
SIGMAS		0.00000	0.00000	0.00000
1	O	0.00000	1.06600	0.19510
SIGMAS		0.00000	0.00040	0.00040
2	O	0.00000	0.43400	0.19510
SIGMAS		0.00000	0.00040	0.00040
3	O	0.00000	0.93400	-0.19510
SIGMAS		0.00000	0.00040	0.00040
4	O	0.00000	0.56600	-0.19510
SIGMAS		0.00000	0.00040	0.00040
5	O	-0.18400	0.75000	0.44510
SIGMAS		0.00040	0.00000	0.00000
6	O	0.18400	0.75000	0.44510
SIGMAS		0.00040	0.00000	0.00040
7	O	-0.31600	0.75000	0.05490
SIGMAS		0.00040	0.00000	0.00040
8	O	0.31600	0.75000	0.05490
SIGMAS		0.00040	0.00000	0.00040

ATOMS		DISTANCES			ANGLE
I	J	0-I	0-J	I-J	I-0-J
1	2	2.12861	2.12861	4.17385	157.286
1	3	2.12861	2.26722	2.49078	68.947
1	4	2.12861	2.26723	4.04325	133.767
1	5	2.12861	2.26722	2.84018	80.430
1	6	2.12861	2.26723	2.84018	80.430
1	7	2.12861	2.12861	3.06812	92.222
1	8	2.12861	2.12861	3.06812	92.222
2	3	2.12861	2.26722	4.04325	133.767
2	4	2.12861	2.26723	2.49078	68.947
2	5	2.12861	2.26722	2.84018	80.430
2	6	2.12861	2.26723	2.84018	80.430
2	7	2.12861	2.12861	3.06812	92.222
2	8	2.12861	2.12861	3.06812	92.222
3	4	2.26722	2.26723	2.43035	64.820
3	5	2.26722	2.26722	4.19618	135.458
3	6	2.26722	2.26723	4.19618	135.458
3	7	2.26722	2.12861	2.84018	80.430
3	8	2.26722	2.12861	2.84018	80.430
4	5	2.26723	2.26722	4.19618	135.458
4	6	2.26723	2.26723	4.19618	135.458
4	7	2.26723	2.12861	2.84018	80.430
4	8	2.26723	2.12861	2.84018	80.430
5	6	2.26722	2.26723	2.43035	64.820
5	7	2.26722	2.12861	2.49078	68.947
5	8	2.26722	2.12861	4.04325	133.767
6	7	2.26723	2.12861	4.04325	133.767
6	8	2.26723	2.12861	2.49078	68.947
7	8	2.12861	2.12861	4.17385	157.286

FACE DEFINED BY POINTS			PLANE NORMAL			AREA
			OX	OY	OZ	
1	3	7	-0.62286	0.73285	-0.27381	3.32216
1	3	8	-0.62286	-0.73285	0.27381	3.32216
1	5	6	0.00000	0.58233	0.81295	3.11947
1	5	7	0.73285	-0.62286	-0.27381	3.32216
1	6	8	0.73285	0.62286	0.27381	3.32216

```
2  4  7     0.62286 0.73285 0.27381        3.32216
2  4  8     0.62286-0.73285-0.27381        3.32216
2  5  6     0.00000 0.58233-0.81295        3.11947
2  5  7    -0.73285-0.62286 0.27381        3.32216
2  6  8    -0.73285 0.62286-0.27381        3.32216
3  4  7    -0.58233 0.00000-0.81295        3.11947
3  4  8    -0.58233 0.00000 0.81295        3.11947
```

POLYHEDRAL VOLUME   19.00444
SIGMA(S)   0.02290

```
CATION:      SI   0.00000   0.25000   0.37500
   SIGMAS         0.00000   0.00000   0.00000
      1       O   0.00000   0.43400   0.19510
   SIGMAS         0.00000   0.00040   0.00000
      2       O   0.00000   0.06600   0.19510
   SIGMAS         0.00000   0.00040   0.00040
      3       O   0.18400   0.25000   0.55490
   SIGMAS         0.00040   0.00000   0.00040
      4       O  -0.18400   0.25000   0.55490
   SIGMAS         0.00040   0.00000   0.00040
```

ATOMS		DISTANCES			ANGLE
I	J	O-I	O-J	I-J	I-O-J
1	2	1.62291	1.62291	2.43035	96.966
1	3	1.62291	1.62291	2.75356	116.063
1	4	1.62291	1.62291	2.75356	116.063
2	3	1.62291	1.62291	2.75356	116.063
2	4	1.62291	1.62291	2.75356	116.063
3	4	1.62291	1.62291	2.43035	96.966

FACE DEFINED			PLANE NORMAL			AREA
BY POINTS			OX	OY	OZ	
1	2	3	-0.87071	0.00000	0.49179	3.00259
1	2	4	-0.87071	0.00000	-0.49179	3.00259
1	3	4	0.00000	-0.87071	-0.49179	3.00259
2	3	4	0.00000	-0.87071	0.49179	3.00259

```
POLYHEDRAL VOLUME      2.11796
   QUAD. ELONG.        1.02369
   ANGLE VARIANCE     97.75781
SIGMA(S)      0.00517      0.00111      0.00000
```

STRUCTURAL VARIATIONS WITH TEMPERATURE, PRESSURE AND COMPOSITION

Chapter 6

Structural Variations with Temperature

CONTENTS

I   Introduction to Part Two 115
II   Effects of Temperature on the Ionic Bond 116
    A   Potential Energy of the Ionic Bond 116
       (1)   Bond energy 116
       (2)   Site energy 117
       (3)   Crystal energy 119
    B   Effects of Temperature on Crystal Energy 119
    C   Accuracy of Thermal Expansion Coefficients 122
III   Dimensional Variation with Temperature 123
    A   Review of Previous Studies 123
       (1)   Fixed structures 123
       (2)   High-temperature crystal structures 123
    B   Empirical Bond Distance–Temperature Relationships 124
    C   Relationships between Polyhedral and Bulk Thermal Expansion . . . 137
IV   Thermal Parameter Variation with Temperature 139
V   Conclusions 142

I. INTRODUCTION TO PART TWO

Part One of this monograph is an introduction to techniques for collecting and analysing data on the variation of crystal structure with temperature, pressure and composition. This first part is a general treatment, applicable to the study of all crystalline materials. Part Two, on results of such crystallographic studies, is constrained to be less general. Most studies of crystal structures at high temperature or pressure have been performed by earth scientists or ceramicists on minerals or mineral-like compounds. These materials are, for the most part, ionic in the sense used by Pauling (1960); that is their atoms may be characterized as cations with positive electrostatic charge, and anions with negative charge. Oxygen-based compounds are by far the most widely studied, because oxygen is the dominant anion

115

in the earth. What follows, therefore, is based primarily on a relatively select group of inorganic, mineral-like materials. Without further research we cannot know whether generalizations developed in these chapters apply to other types of solids.

In the broadest sense, study of the variation of structure with temperature, pressure and composition encompasses both crystal chemistry and solid-state phase equilibria, for changes in intensive variables may result in either continuous or discontinuous variations in structure. Chapters Six to Nine contain data on the continuous variations in structure resulting from continuous changes of temperature, pressure or composition. Discontinuities or phase transitions, which may occur as a consequence of geometrical limits to structure, are considered in Chapter Ten.

A recurrent theme of Part Two is the systematic and predictable nature of structural variations in ionic crystals with changes in temperature, pressure and composition. Our approach to the description and analysis of these systematic variations is primarily empirical, although we have also attempted to present a rationale for crystal behaviour by invoking a simple two-term Coulombic bonding model. The two-term bonding model, although simplistic, is useful in gaining an intuitive feeling for physical causes of structural variations. More sophisticated and realistic models of bonding are necessary, however, if crystalline properties are to be predicted from first principles. It is hoped that the empirical relationships developed in the following pages may help in establishing boundary conditions for those more rigorous bonding models.

II. EFFECTS OF TEMPERATURE ON THE IONIC BOND

A. Potential energy of the ionic bond

(1) Bond energy

The ionic bond is a consequence of electrostatic attraction between ions of opposite charge. The bond energy between a cation and anion, defined as the energy required to remove the ions to an infinite separation from their equilibrium positions, is thus primarily a function of the charges and separation of the two ions. In the simplest model, bond energy is assumed to arise from only two terms, the Coulombic or electrostatic attraction potential and the repulsive potential. Smaller energy contributions from Van der Waal's interactions, polarization, incomplete ionization of the atoms and atomic vibrations generally account for less than 2 per cent of the total cohesive energy in ionic bonds, and thus are ignored in this first-approximation treatment. If the energy of the ions is chosen to be zero at infinite separation, then the bond energy will be negative at equilibrium separation.

The electrostatic potential of an ion pair is given by:

$$U_A = z_c z_a e^2 / d \qquad (6\text{-}1)$$

where d is the cation–anion separation, e is the charge on an electron, and z_c and z_a are cation and anion charges, respectively. The repulsive energy term may be given in either of two forms. The Born exponential form is commonly cited:

$$U_R = \lambda \exp(-d/\rho) \tag{6-2}$$

Alternatively, the repulsive potential may be represented by a power-law term of the form:

$$U_R = B/d^n \tag{6-3}$$

The choice between equations (6-2) and (6-3) is often based on computational convenience; neither form of the repulsive term has any physical basis. The repulsive term is due primarily to inner electron orbital interactions: as electron clouds overlap some electrons are forced into higher energy states because of restrictions related to Pauli's exclusion principle. Thus, in the region of orbital interpenetration, the repulsive force increases sharply as interatomic separation decreases. The parameters λ and ρ (equation 6-2) or B and n (equation 6-3) are constants that depend on the electronic structure of the two ions.

The total bond energy may be given by the sum of equations (6-1) and (6-3):

$$U_{total} = z_c z_a e^2/d + B/d^n \tag{6-4}$$

Calculated curves of bond potential energy versus interatomic separation show a characteristic 'well' shape, or potential minimum, that represents equilibrium separation between ions (see e.g. Figure 6-1).

(2) Site energy

Site energy, which is closely related to bond energy, is defined as the energy required to separate a particular ion to an infinite separation from its equilibrium position in a crystal. Several authors, including Pauling (1960), Kittel (1971) and Ohashi and Burnham (1972) have presented procedures for calculating site energy with a two-term potential. The electrostatic site energy of the jth ion is the sum of all Coulomb potentials from other ions to the jth ion:

$$U_A(j) = z_j \sum_h \sum_k \sum_l \sum_i^m \frac{z_i e^2}{d_{ij(hkl)}} \tag{6-5}$$

where $d_{ij(hkl)}$ is the distance between the jth ion in the origin unit cell and the ith ion in cell hkl, z_j and z_i are ion valences, and m is the number of atoms per unit cell. Note that the 'self' term, between the jth ion in the origin unit cell and itself, must be excluded from the summation. Equation (6-5) is often presented in an abbreviated form by introducing the Madlung constant, A_j:

$$A_j = z_j d_j \sum_h \sum_l \sum_i^m \frac{z_i}{d_{ij(hkl)}} \tag{6-6}$$

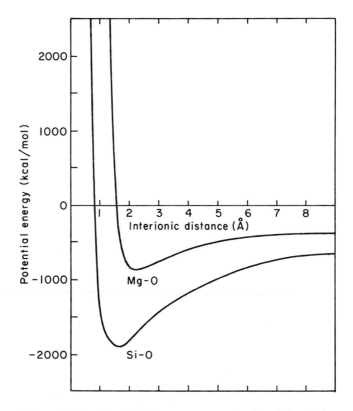

Figure 6-1. Bond potential energy curves for Mg–O bonds in MgO (periclase) and Si–O bonds in α-SiO$_2$ (quartz). Curves are calculated from equation (6-9). For MgO, $A = 6.990$, $B = 1.230 \times 10^{-21}$ kcal/mol, and $n = 5.450$. For SiO$_2$, $A = 17.75$, $B = 3.017 \times 10^5$ kcal/mol, and $n = 2.104$. (After Hazen, 1975)

where d_j is the nearest-neighbour distance to the jth ion. The Madlung constant is dimensionless, and is a function of the spatial distribution of ions about a given ion site. A different Madlung constant must be evaluated for each site geometry. Thus, whereas all alkali halides with the NaCl structure have the same Madlung constant, structures of lower symmetry with variable atomic positions may have Madlung constants that vary continuously with temperature, pressure and composition.

Site repulsive energy is simply the sum of bond repulsive energies of all N nearest neighbours to the jth ion:

$$U_R(j) = \sum_{i}^{N} B_{ij}/d_{ij}^{n_j} \tag{6-7}$$

If all nearest-neighbour bonds to the jth ion are identical, as in the NaCl structure, then equation (6-7) reduces to:

$$U_R(j) = N(B_{ij}/d_{ij}^{n_j})$$
(6-8)

Total site energy is the sum of equations (6-5) and (6-7):

$$U_{total}(j) = A_j e^2/d_j + \sum_i^N B_{ij}/d_{ij}^{n_j}$$
(6-9)

Equation (6-9) may be used to calculate site potential curves in the graphical form of energy versus cation–anion separation, as illustrated in Figure 6-1. In addition to providing a useful visualization of the ionic bond, site energy calculations have proven successful in rationalizing the distribution of cations among several non-equivalent crystallographic sites in complex oxides (Ohashi and Burnham, 1972).

(3)   Crystal energy

Crystal energy is defined as the energy required to remove *all* ions in a crystal to an infinite separation from their equilibrium positions. Thus the crystal energy is one half the sum of all site energies:

$$U_{crystal} = 1/2 \sum_i U(i)$$
(6-10)

Crystal energy calculations (e.g. Kittel, 1971, Table 5), in spite of the simplicity of the two-term model, have proved successful in estimating the cohesive energy of some ionic solids. In the case of many compounds, including the alkali halides, the relative stability of two different possible structures may be predicted using equation (6-10). In addition, as will be demonstrated in the subsequent sections of this monograph, the two-term model of bond, site and crystal energies may be used to understand qualitative aspects of the effects of temperature, pressure and composition on the ionic bond.

B.   Effects of temperature on crystal energy

The addition of heat to an ionic crystal increases the energy of the crystal, primarily in the form of lattice vibrations or phonons, manifest in the oscillation of ions or groups of ions. When ionic bonds are treated as classical harmonic oscillators, the principal calculated effect of temperature is simply increased vibration amplitude, with eventual breakage of bonds at high temperature due to extreme amplitudes. This model is useful in rationalizing phenomena such as melting, site disordering, or increased electrical conductivity at high temperature.

The purely harmonic model of atomic vibrations is not adequate to explain many properties of crystals, however, and anharmonic vibration terms must be considered in any analysis of the effect of temperature on crystal structure. A mathematical treatment of anharmonic vibrations is beyond the scope of this work, but an introduction is provided by Kittel (1971). An important consequence of anharmonic motion is the change in equilibrium bond distance with temperature, i.e. thermal expansion. Dimensional changes of a crystal structure with temperature may be represented by the coefficient of thermal expansion, α, defined as:

$$\text{linear } \alpha_l = 1/d(\partial d/\partial T)_{P,X} \tag{6-11}$$

$$\text{volume } \alpha_v = 1/V(\partial V/\partial T)_{P,X} \tag{6-12}$$

where subscripts P and X denote partials at constant pressure and composition, respectively. Another useful measure is the mean coefficient of expansion between two temperatures, T_1 and T_2:

$$\text{mean } \bar{\alpha}_{T_1,T_2} = 2/(d_1 + d_2)\left[\frac{(d_2 - d_1)}{(T_2 - T_1)}\right] \cong \alpha_{(T_1+T_2)/2} \tag{6-13}$$

The mean coefficient of thermal expansion is the most commonly reported parameter in experimental studies of structure variation with temperature.

A useful intuitive approach to understanding thermal expansion stems from the fact that the potential diagram of an ionic bond, with potential energy plotted as a function of inter-ionic separation (e.g. Figure 6-1), is asymmetric about the minimum potential. Thus, as the potential energy of a system increases with the addition of heat, the equilibrium separation of ions changes in proportion to the asymmetry (Figure 6-2). Note that there is no constraint on the asymmetry of the potential well, and thermal expansion may be either positive or negative. There is no simple functional form that successfully models thermal expansion in all materials. For want of a more satisfactory theoretically based equation, most volume thermal expansion data are presented as a simple second-order polynomial:

$$V = V_0 + aT + bT^2 \tag{6-14}$$

The coefficient of thermal expansion is a function of temperature as illustrated in Figure 6-3. Near absolute zero, because there is virtually no change in potential energy of a system with temperature, there is also little thermal expansion of bonds. In fact, a small range of negative thermal expansion has been observed in some compounds below 30 K (White, 1973). As the potential energy increases, so does thermal expansion. Thus, as proposed by Debye, there exists a close relationship between thermal expansion and the specific heat, C_v. In the case of a cubic crystal with a single oscillation frequency, the relationship between linear thermal expansion coefficient and specific heat is particularly straightforward:

$$\alpha_l = 1/2 \left(\frac{n + 2}{n}\right) C_v/U \tag{6-15}$$

where n is the repulsive exponent from equation (6-3) and U is the crystal energy of equation (6-10). Megaw (1938), Kumar (1959) and Das et al. (1963) are among the authors who have used equation (6-15) and related expressions to calculate thermal expansion.

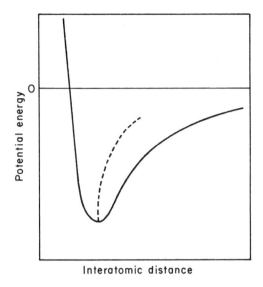

Figure 6-2. Equilibrium interatomic distance of a diatomic oscillator increases as a function of temperature because of the asymmetry of the potential well

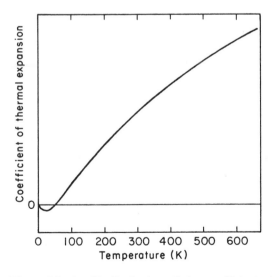

Figure 6-3. An idealized plot of the coefficient of thermal expansion versus temperature. Thermal expansion is zero at 0 K. An interval of negative α is sometimes observed below 30 K (White, 1973)

C. Accuracy of thermal expansion coefficients

Much of the description of structural variation with temperature depends on the use of thermal expansion coefficients of volume and linear structural units. It is important, therefore, to recognize the limits in the accuracy of reported coefficients. Unit-cell expansion coefficients may be determined on single crystals or powders, using either x-ray diffraction or dilatometry. Although the reported precision for many of these experiments exceeds 1 per cent in the expansion coefficient, several studies on the same material commonly differ by ± 10 per cent.

A variety of factors may lead to the disagreement between thermal expansion studies. In some materials impurities, crystalline defects, or sample preparation may play a major role. However, even in well crystallized pure compounds, such as MgO or Al_2O_3, there exists a considerable range of published expansion coefficients. It appears, therefore, that systematic experimental errors are very common in thermal expansion measurements. Evidently standardized procedures are needed to increase the accuracy of these studies. Until that time, the accuracy of any given study must be conservatively accepted as no better than ± 5 per cent, even though reported precision may be much smaller.

Errors in bond distance expansion coefficients must be significantly greater than 5 per cent because of the unknown effects of correlated thermal motion. As noted in Chapter 5 (section IV-A-(1)), an important aspect of bond distance analysis, especially when comparing distances at different temperatures, is the effect of thermal motion on mean interatomic separation, which is the instantaneous separation averaged over time. Atomic fractional coordinates from crystal structure refinements represent the centre of electron density arising from both static (positional) and dynamic (vibrational) effects. Interatomic distances reported in most studies, including this one, are calculated as the distance between two such centres. Busing and Levy (1964), however, have demonstrated that a better measure of interatomic distance is the *mean* separation. In general, when thermal vibrations are large, the mean separation of two atoms will be greater than the separation of mean atomic position (Figure 5-1). Thus, thermal expansion of a bond based on mean separation may be greater, and may represent a more valid physical situation, than expansions based on distances between atomic centres, which are reported in most recent studies.

Consider the case of Al–O bond expansion in $NaAlSi_3O_8$, as illustrated in Figure 5-2. The thermal expansion coefficient is zero or perhaps even slightly negative when calculated in the conventional way, with uncorrected bond distances. A similar conclusion would result from thermal expansion calculations based on purely correlated atomic motions. In the other extreme with anticorrelated thermal motion, however, the Al–O expansion coefficient is approximately 4×10^{-5} °C^{-1}, which is several times *larger* than the observed linear thermal expansion of most oxides and silicates. Somewhat correlated atomic motions are expected for ionically bonded atoms, so the use of uncorrected distances in thermal expansion calculations may provide a reasonable model for comparing the temperature variation of different bonds. It is important to remember, however, that in the subsequent discussion of bond thermal expansion the important effects of thermal motions have not been included explicitly because of the lack of data on atomic motion correlations. A complete understanding of bond expansion must await this additional information.

III. DIMENSIONAL VARIATION WITH TEMPERATURE

A. Review of previous studies

Data on the variation of structural dimensions with temperature or pressure come from two distinct types of studies. Complete three-dimensional structure refinements are the most obvious sources of data, but it is also possible to derive this information from unit-cell dimensions alone in the case of many constrained or simple structures. Data from these two sources are reviewed below.

(1) Fixed structures

NaCl, CsCl, CaF_2 and cubic ZnS are all common fixed structures in which there are no variable positional parameters. Thermal expansion data on materials that crystallize in these structures thus provide information on bond thermal expansion as well. Other simple structures, including those of rutile (TiO_2), corundum (Al_2O_3), hexagonal ZnS, and ZrO_2 also have bulk expansions that are similar to expansions of cation–anion bonds (see below). A partial list of linear expansion coefficients for these fixed- and simple-structure compounds, together with bonding parameters, is given in Table 6-1. For uniformity, the bond expansion coefficients are calculated as the mean expansion from 23°C to 1000°C. For most compounds studied the variation of α with temperature is linear. And a useful approximation for compounds not yet studied to 1000°C is $\bar{\alpha}_{1000} \approx \alpha_{510}$. The assumed accuracy of bond expansion coefficients is no greater than ± 10 per cent, as described above and as indicated by the significant inconsistencies in reported thermal expansion coefficients for specific compounds (MgO, Al_2O_3, CaO) by different authors.

(2) High-temperature crystal structures

Several dozen high-temperature structure studies provide a wealth of data on the variation of cation–anion (primarily cation–oxygen) bonds with temperature. A list of high-temperature structure refinements and the observed coefficients of bond expansion (as defined above) are tabulated along with thermal and bonding parameters in Table 6-2. Bond expansion coefficients are calculated from weighted linear regression of reported bond distance versus temperature data; estimated standard errors on the coefficients are assumed to be ± 10 per cent or greater as noted above.

B. Empirical bond distance–temperature relationships

Several attempts have been made by previous investigators to relate thermal expansion to bonding parameters or other physical properties. Megaw (1938) proposed that in crystals with only one type of bond (homodesmic crystals) thermal expansion is a function of Pauling bond strength (defined as z/n, where z and n are cation valence and coordination number, respectively). She proposed the relationship:

$$\alpha \propto n^2/z^2 \qquad (6\text{-}16)$$

This equation is in good agreement with the observed thermal expansion of many simple compounds, but it fails for many of the bonds listed in Tables 6-1 and 6-2, especially in the case of alkali metal–oxygen bonds.

In a more recent study of pyroxene crystal structures at high temperature, Cameron et al. (1973) related expansion coefficients of metal–oxygen bonds to bond strengths (defined as $4v^2\mu\pi^2c^2$, where v is M–O stretching frequency, μ is reduced mass, and c is the velocity of light). Although their relationship successfully modelled the bond expansion of their study, the Cameron et al. equation does not predict many of the features of Tables 6-1 and 6-2, such as the similar expansion coefficients of Ni–O, Mg–O, Fe–O and Ba–O in rock-salt type structures.

Some authors have attempted to relate thermal expansivity to other physical variables. Hanneman and Gatos (1965) demonstrated that thermal expansion and compressibility are proportional for cubic metals and alloys. Unique and continuous functions exist for each cubic metal structure type. Van Uitert and coworkers (1977a, 1977b) found a simple inverse relationship between coefficient of thermal expansion and melting temperature for a large number of close-packed structures. Many investigators have examined the interrelationships between thermal expansion, thermal vibration amplitudes and specific heat (e.g. Lonsdale, 1959; Kumar, 1959; Deganello, 1978). The number of attempts to predict thermal expansion and relate expansion to other physical variables attests to the usefulness of such relationships in modelling the high-temperature behaviour of the solid state.

Data in Tables 6-1 and 6-2 lead to several simple empirical relationships, which enable the prediction of bond distance variation with temperature. The first important observation is that all cation coordination polyhedra of a given type (i.e. all magnesium–oxygen octahedra or silicon–oxygen tetrahedra) demonstrate similar expansion coefficients. For example, Tables 6-1 and 6-2 contain eighteen different magnesium octahedra plus two magnesium–(oxygen, fluorine) octahedra. All of these Mg–O thermal expansion coefficients are consistent with a value of $14 \times 10^{-6}\,°C^{-1}$ within one estimated standard deviation (Figure 6-4).

Of 48 silicon tetrahedra in Tables 6-1 and 6-2, 43 have expansion coefficients near or equal to zero within two standard errors. Of the remaining five, two tetrahedra have positive expansion coefficients. In anorthite the positive expansion of Si–O is accompanied by a large *contraction* of adjacent Al–O bonds. It

Table 6-1. Thermal expansion of fixed-structure and simple-structure compounds for which bond thermal expansion is equal to bulk linear expansion

Compound	Bond	Structure Type	z_c	z_a	$d(\text{Å})$	n	$\bar{\alpha}^*_{1000} \times 10^{-6}$ (K^{-1})	Reference
NiO	Ni–O	NaCl	2	2	2.08	6	13.5	Nielsen and Leipold (1965)
NiO	Ni–O	NaCl	2	2	2.08	6	16.7	Bobrovskii et al. (1973)
MgO	Mg–O	NaCl	2	2	2.10	6	13.8	Austin (1931)
MgO	Mg–O	NaCl	2	2	2.10	6	12.7	Skinner (1957)
MgO	Mg–O	NaCl	2	2	2.10	6	13.3	Beals and Cook (1957)
MgO	Mg–O	NaCl	2	2	2.10	6	12.7	Suzuki (1975)
MgO	Mg–O	NaCl	2	2	2.10	6	12.4	Hazen (1976a)
$(Mg_{.33}Fe_{.67})O$	(Mg, Fe)–O	NaCl	2	2	2.14	6	14.4	Carter (1959)
$(Mg_{.37}Fe_{.63})O$	(Mg, Fe)–O	NaCl	2	2	2.14	6	13.1	Rigby et al. (1946)
$(Mg_{.64}Fe_{.35})O$	(Mg, Fe)–O	NaCl	2	2	2.12	6	13.5	Rigby et al. (1946)
$(Mg_{.84}Fe_{.16})O$	(Mg, Fe)–O	NaCl	2	2	2.11	6	13.5	Rigby et al. (1946)
FeO	Fe–O	NaCl	2	2	2.16	6	12.2	Rigby et al. (1946)
FeO	Fe–O	NaCl	2	2	2.16	6	15.2	Carter (1959)
CoO	Co–O	NaCl	2	2	2.13	6	13.8	Min'ko (1972)
MnO	Mn–O	NaCl	2	2	2.22	6	14.1	Brooksbank and Andrews (1968)
CdO	Cd–O	NaCl	2	2	2.35	6	13.4	Valeev and Kvaskov (1973)
CdO	Cd–O	NaCl	2	2	2.35	6	13.2	Singh and Dayal (1969)
CaO	Ca–O	NaCl	2	2	2.41	6	13.0	Beals and Cook (1957)
CaO	Ca–O	NaCl	2	2	2.41	6	13.6	Grain and Campbell (1962)
SrO	Sr–O	NaCl	2	2	2.58	6	13.7	Beals and Cook (1957)
BaO	Ba–O	NaCl	2	2	2.77	6	17.8	Eisenstein (1946)
BaO	Ba–O	NaCl	2	2	2.77	6	12.8	Zollweg (1955)

Table 6-1 *continued*

Compound	Bond	Structure Type	z_c	z_a	$d(\text{Å})$	n	$\bar{\alpha}^*_{1000} \times 10^{-6}$ (K^{-1})	Reference
BeO	Be–O	Zincite	2	2	1.66	4	8.4	Beals and Cook (1957)
BeO	Be–O	Zincite	2	2	1.66	4	9.1	Grain and Campbell (1962)
ZnO	Zn–O	Zincite	2	2	1.80	4	6.8	Beals and Cook (1957)
ZnO	Zn–O	Zincite	2	2	1.80	4	7.7	Valeev and Kvaskov (1973)
Al_2O_3	Al–O	Corundum	3	2	1.91	6	8.4	Austin (1931)
Al_2O_3	Al–O	Corundum	3	2	1.91	6	9.3	Beals and Cook (1957)
Al_2O_3	Al–O	Corundum	3	2	1.91	6	8.6	Campbell and Grain (1961)
Al_2O_3	Al–O	Corundum	3	2	1.91	6	8.1	Petukhov (1973)
Fe_2O_3	Fe–O	Corundum	3	2	2.03	6	10.3	Sharma (1950)
Bi_2O_3	Bi–O	Bismite	3	2	2.44	6	9.4	Valeev and Kvaskov (1973)
Cr_2O_3	Cr–O	Corundum	3	2	1.99	6	7.2	Rigby *et al.* (1946)
Cr_2O_3	Cr–O	Corundum	3	2	1.99	6	7.9	Brookshank and Andrews (1968)
UO_2	U–O	Fluorite	4	2	2.37	8	10.6	Kempter and Elliott (1959)
ThO_2	Th–O	Fluorite	4	2	2.42	8	9.3	Kempter and Elliott (1959)
ThO_2	Th–O	Fluorite	4	2	2.42	8	8.2	Skinner (1957)
ThO_2	Th–O	Fluorite	4	2	2.42	8	9.6	Grain and Campbell (1962)
CeO_2	Ce–O	Fluorite	4	2	2.34	8	12.6	Stecura and Campbell (1962)
ZrO_2	Zr–O	Baddeleyite	4	2	2.16	7	8.0	Austin (1931)
ZrO_2	Zr–O	Baddeleyite	4	2	2.16	7	8.4	Min'ko (1972)
HfO_2	Hf–O	Baddeleyite	4	2	2.17	7	7.1	Grain and Campbell (1962)
β-SiO_2	Si–O	β-Quartz	4	2	1.61	4	0.0	Ackermann and Sorrell (1974)
ReO_3	Re–O	ReO_3	6	2	1.87	6	1.1	Chang and Trucano (1978)
LiF	Li–F	NaCl	1	1	2.02	6	46	Skinner (1966)
NaF	Na–F	NaCl	1	1	2.31	6	45	Pathak *et al.* (1973)

							Reference
NaCl	Na–Cl	NaCl	1	2.82	6	51	Skinner (1966)
NaCl	Na–Cl	NaCl	1	2.82	6	55	Pathak and Vasavada (1970)
KCl	K–Cl	NaCl	1	3.15	6	46	Pathak and Vasavada (1970)
KCl	K–Cl	NaCl	1	3.15	6	51	Pathak and Vasavada (1970)
K·Br	K–Br	NaCl	1	3.26	6	45	Pathak et al. (1973)
KBr	K–Br	NaCl	1	3.26	6	49	Skinner (1966)
RbBr	RB–Br	NaCl	1	3.43	6	44	Pathak et al. (1973)
KI	K–I	NaCl	1	3.53	6	47	Skinner (1966)
CsBr	Cs–Br	CsCl	1	3.71	8	68	Pathak and Vasavada (1970)
CaF$_2$	Ca–F	CaF$_2$	2	2.36	8	21	Kumar (1959)
CaF$_2$	Ca–F	CaF$_2$	2	2.35	8	22	Larionov and Malkin (1975)
MnS	Mn–S	NaCl	2	2.61	6	18	Brooksbank and Andrews (1968)
PbS	Pb–S	NaCl	2	2.97	6	22	Skinner (1962)
ZnS	Zn–S	Cubic ZnS	2	2.34	4	9	Skinner (1962)
ZnS	Zn–S	Hex. ZnS	2	2.24	4	10	Skinner (1962)
PbTe	Pb–Te	NaCl	2	3.23	6	20	Houston et al. (1968)
PbSe	Pb–Se	NaCl	2	3.06	6	21	Skinner (1962)
ZnSe	Zn–Se	Cubic ZnS	2	2.45	4	9	Skinner (1962)
AlAs	Al–As	Cubic ZnS	3	2.43	4	5.3	Ettenberg and Paff (1970)
GaAs	Ga–As	Cubic ZnS	3	2.45	4	6.7	Feder and Light (1968)
TiN	Ti–N	NaCl	4	2.12	6	9.4	Brooksbank and Andrews (1968)
BN	B–N	Cubic ZnS	3	1.45	4	13.0	Pease (1952)
UN	U–N	NaCl	4	2.44	6	8.6	Kempter and Elliott (1959)
NbC	Nb–C	NaCl	4	2.23	6	7	Jun (1970)
TaC	Ta–C	NaCl	4	2.23	6	7	Jun (1970)
TiC	Ti–C	NaCl	4	2.16	6	8	Skinner (1966)
ZrC	Zr–C	NaCl	4	2.34	6	7	Skinner (1966)
C	C–C	Diamond	4	1.54	4	3.5	Skinner (1957)

$$^*\bar{a}_{1000} = \frac{2}{d_0 + d_{1000}} \left(\frac{d_{1000} - d_0}{980} \right) \approx a_{510},$$

where d_0 and d_{1000} are mean cation–anion bond distance at 20°C and 1000°C, respectively.

Table 6.2. Polyhedral thermal expansion, variation of isotropic temperature parameters and bonding parameters from complete three-dimensional, high-temperature ($\geq 400°C$) crystal structure refinements

Bond	Structure	Site	Formula	Mineral Name	z_c	z_a	$d(\text{Å})$	n	$\bar{\alpha}_{1000} \times 10^6$ (K^{-1})	$\partial B/\partial T$ $(\text{Å}^2/\text{K})$	Reference
Mg–O	NaCl		MgO	Periclase	2	2	2.106	6	12.4(1)	0.0017(1)	Hazen (1976a)
V–O	Corundum		V_2O_3	Karelianite	3	2	2.010	6	13(1)	0.0008(1)	Robinson (1975)
Ti–O	Corundum		Ti_2O_3		3	2	2.046	6	8(1)	0.0010(1)	Rice and Robinson (1977)
Si–O	Cristobalite		SiO_2	Cristobalite	4	2	1.609	4	0(4)	0.0007(4)	Peacor (1973)
Ti–O	Rutile		TiO_2	Rutile	4	2	1.959	6	8(1)	0.0014(1)	Meagher and Lager (1979)
Ti–O	Brookite		TiO_2	Brookite	4	2	1.960	6	6(2)	0.0014(1)	Meagher and Lager (1979)
Ti–O	Anatase		TiO_2	Anatase	4	2	1.949	6	8(1)	0.0016(1)	Horn et al. (1972)
Mg–O	Garnet		$Mg_3Al_2Si_3O_{12}$	Pyrope	2	2	2.269	8	13(1)	0.0020(1)	Meagher (1975)
Al–O	Garnet				3	2	1.887	6	7(1)	0.00092(2)	Meagher (1975)
Si–O					4	2	1.635	4	2(2)	0.0006(1)	Meagher (1975)
Ca–O	Garnet		$Ca_3Al_2Si_3O_{12}$	Grossular	2	2	2.406	8	10(1)	0.0012(1)	Meagher (1975)
Al–O	Garnet				3	2	1.921	6	10(1)	0.0008(1)	Meagher (1975)
Si–O					4	2	1.647	4	6(2)	0.0010(3)	Meagher (1975)
Mg–O	Olivine	M1	Mg_2SiO_4	Forsterite	2	2	2.095	6	16(3)	0.0015(1)	Hazen (1976b)
Mg–O		M2			2	2	2.133	6	16(2)	0.0015(1)	Hazen (1976b)
Si–O		T			4	2	1.630	4	–1(3)	0.0009(1)	Hazen (1976b)
(Mg, Fe)–O	Olivine	M1	$(Mg_{1.4}Fe_{0.6})SiO_4$	Hortonolite	2	2	2.118	6	12(1)	0.0016(1)	Brown and Prewitt (1973)
(Mg, Fe)–O		M2			2	2	2.148	6	12(1)	0.0014(1)	Brown and Prewitt (1973)
Si–O		T			4	2	1.638	4	–1(1)	0.0009(1)	Brown and Prewitt (1973)
(Fe, Mg)–O	Olivine	M1	$(Mg_{0.75}Fe_{1.10}Mn_{0.15})SiO_4$	Hortonolite	2	2	2.135	6	16(4)	0.0015(1)	Smyth and Hazen (1973)
(Fe, Mg)–O		M2			2	2	2.167	6	14(3)	0.0015(1)	Smyth and Hazen (1973)
Si–O		T			4	2	1.629	4	2(3)	0.0009(1)	Smyth and Hazen (1973)
Fe–O	Olivine	M1	Fe_2SiO_4	Fayalite	2	2	2.157	6	12(1)	0.0032(2)	Smyth (1975)
Fe–O		M2			2	2	2.179	6	14(1)	0.0028(2)	Smyth (1975)
Si–O		T			4	2	1.628	4	–4(3)	0.0018(2)	Smyth (1975)

Structure	Mineral	Formula	Site	Bond						Reference
Olivine	Ni-olivine	Ni_2SiO_4	M1	Ni–O	2	2.078	6	15(1)	0.0012(1)	Lager and Meagher (1978)
			M2	Ni–O	2	2.100	6	13(1)	0.0013(1)	Lager and Meagher (1978)
			T	Si–O	4	1.639	4	0(1)	0.0008(1)	Lager and Meagher (1978)
Olivine	Monticellite	$CaMgSiO_4$	M1	Ca–O	2	2.129	6	18(1)	0.0020(2)	Lager and Meagher (1978)
			M2	(Mg, Fe)–O	2	2.368	6	13(1)	0.0019(2)	Lager and Meagher (1978)
			T	Si–O	4	1.637	4	–3(3)	0.0015(1)	Lager and Meagher (1978)
Olivine	Glaucochroite	$Ca(Mn_{0.85}Mg_{0.10}Zn_{0.05})SiO_4$	M1	Ca–O	2	2.210	6	15(1)	0.0025(1)	Lager and Meagher (1978)
			M2	(Mn, Mg)–O	2	2.366	6	15(1)	0.0019(1)	Lager and Meagher (1978)
			T	Si–O	4	1.640	4	–7(3)	0.0017(1)	Lager and Meagher (1978)
Spinel	Ni-olivine	Ni_2SiO_4	M	Ni–O	2	2.060	6	13(1)	0.0012(1)	Finger et al. (1979)
			T	Si–O	4	1.660	4	1(4)	0.0009(1)	Finger et al. (1979)
Sphene	Sphene	$CaTiSiO_5$		Ca–O	2	2.450	7	16(3)	0.0024(1)	Taylor and Brown (1976)
				Ti–O	4	1.956	6	9(2)	0.0012(2)	Taylor and Brown (1976)
				Si–O	4	1.639	4	0(2)	0.0007(1)	Taylor and Brown (1976)
Pseudobrookite	Armalcolite	$(Fe, Mg)Ti_2O_5$		(Fe, Mg)–O	2	2.103	6	–11(3) †	0.0014(1)	Wechsler (1977)
				Ti–O	4	1.979	6	22(2)	0.0014(1)	Wechsler (1977)
Perovskite		$PbTiO_3$		Pb–O	2	2.843	12	23(8) ‡	0.0038(2)	Glazer and Mabud (1978)
				Ti–O	4	2.013	6	20(10)	0.0012(6)	Glazer and Mabud (1978)
Orthopyroxene	Orthoferrosilite	$FeSiO_3$	M1	Fe–O	2	2.135	6	16(1)	0.0018(1)	Sueno et al. (1976)
			M2	Fe–O	2	2.352	7	24(2) §	0.0025(1)	Sueno et al. (1976)
			SiA	Si–O	4	1.626	4	–3(1)	0.0012(1)	Sueno et al. (1976)
			SiB	Si–O	4	1.637	4	–7(1)	0.0012(1)	Sueno et al. (1976)
Orthopyroxene	Ferrohypersthene	$(Mg_{.3}Fe_{.7})SiO_3$	M1	(Mg, Fe)–O	2	2.104	6	20(2)	0.0020(1)	Smyth (1973)
			M2	Fe–O	2	2.340	7	18(2) §	0.0030(1)	Smyth (1973)
			SiA	Si–O	4	1.628	4	–3(1)	0.0018(1)	Smyth (1973)
			SiB	Si–O	4	1.637	4	–16(2)	0.0019(1)	Smyth (1973)
Clinopyroxene (2-chain)	Clinohypersthene	$(Mg_{.3}Fe_{.7})SiO_3$	M1	(Mg, Fe)–O	2	2.111	6	16(2) †	0.0020(2)	Smyth (1974)
			M2	Fe–O	2	2.297	7	–2(4) §	0.0026(1)	Smyth (1974)
			SiA	Si–O	4	1.625	4	–5(5)	0.0015(2)	Smyth (1974)
			SiB	Si–O	4	1.640	4	–24(4)	0.0020(2)	Smyth (1974)

Table 6.2 continued

Bond	Structure	Site	Formula	Mineral Name	z_c	z_a	$d(Å)$	n	$\bar{\alpha}^*_{1000} \times 10^6$ (K^{-1})	$\partial B/\partial T$ $(Å^2/K)$	Reference
Al–O	Clinopyroxene	M1	$LiAlSi_2O_6$	Spodumene	3	2	1.919	6	10(1)	0.0012(1)	Cameron et al. (1973)
Li–O		M2			1	2	2.211	6	20(1)	0.0041(1)	Cameron et al. (1973)
Si–O		T			4	2	1.618	4	1(1)	0.0011(1)	Cameron et al. (1973)
Fe–O	Clinopyroxene	M1	$NaFe^{3+}Si_2O_6$	Acmite	3	2	2.025	6	8(1)	0.0012(1)	Cameron et al. (1973)
Na–O		M2			1	2	2.518	8	13(1)	0.0034(1)	Cameron et al. (1973)
SiO		T			4	2	1.628	4	1(1)	0.0011(1)	Cameron et al. (1973)
Al–O	Clinopyroxene	M1	$NaAlSi_2O_6$	Jadeite	3	2	1.929	6	10(1)	0.0009(1)	Cameron et al. (1973)
Na–O		M2			1	2	2.469	8	13(1)	0.0026(1)	Cameron et al. (1973)
Si–O		T			4	2	1.625	4	1(1)	0.0008(1)	Cameron et al. (1973)
Cr–O	Clinopyroxene	M1	$NaCrSi_2O_6$	Ureyite	3	2	1.988	6	6(1)	0.0008(1)	Cameron et al. (1973)
Na–O		M2			1	2	2.489	8	13(1)	0.0029(1)	Cameron et al. (1973)
Si–O		T			4	2	1.624	4	2(2)	0.0009(1)	Cameron et al. (1973)
Fe–O	Clinopyroxene	M1	$CaFeSi_2O_6$	Hedenbergite	2	2	2.130	6	10(1)	0.0015(1)	Cameron et al. (1973)
Ca–O		M2			2	2	2.511	8	16(1)	0.0022(1)	Cameron et al. (1973)
Si–O		T			4	2	1.635	4	0(1)	0.0010(1)	Cameron et al. (1973)
Mg–O	Clinopyroxene	M1	$CaMgSi_2O_6$	Diopside	2	2	2.077	6	14(1)	0.0016(1)	Cameron et al. (1973)
Ca–O		M2			2	2	2.498	8	16(1)	0.0023(1)	Cameron et al. (1973)
Si–O		T			4	2	1.635	4	1(1)	0.0011(1)	Cameron et al. (1973)
Na–O	Feldspar	Na	$NaAlSi_3O_8$	High albite	1	2	2.807	9	17(1)	0.0046(2)	Prewitt et al. (1976)
$(Al_{.25}Si_{.75})$–O		T_{1o}			3.75	2	1.646	4	–1(1)	0.0015(1)	Prewitt et al. (1976)
$(Al_{.25}Si_{.75})$–O		T_{1m}			3.75	2	1.641	4	1(1)	0.0015(1)	Prewitt et al. (1976)
$(Al_{.25}Si_{.75})$–O		T_{2o}			3.75	2	1.641	4	–3(1)	0.0015(1)	Prewitt et al. (1976)
$(Al_{.25}Si_{.75})$–O		T_{2m}			3.75	2	1.642	4	–4(1)	0.0015(1)	Prewitt et al. (1976)
Na–O	Feldspar	Na	$NaAlSi_3O_8$	High albite	1	2	2.807	9	18(1)	0.0045(1)	Winter et al. (1979)
$(Al_{.25}Si_{.75})$–O		T_{1o}			3.75	2	1.649	4	–3(1)	0.0015(1)	Winter et al. (1979)
$(Al_{.25}Si_{.75})$–O		T_{1m}			3.75	2	1.642	4	0(1)	0.0015(1)	Winter et al. (1979)
$(Al_{.25}Si_{.75})$–O		T_{2o}			3.75	2	1.640	4	0(1)	0.0015(1)	Winter et al. (1979)
$(Al_{.25}Si_{.75})$–O		T_{2m}			3.75	2	1.642	4	–1(1)	0.0015(1)	Winter et al. (1979)

Group	Formula		Site	Bond							Reference
Feldspar	NaAlSi$_3$O$_8$	Low albite	Na	Na–O	1	2	2.634	7	35(2)	0.0082(1)	Winter et al. (1977)
			T$_{1o}$	Al–O	3	2	1.740	4	0(1)	0.0017(1)	Winter et al. (1977)
			T$_{1m}$	Si–O	4	2	1.609	4	–2(1)	0.0015(1)	Winter et al. (1977)
			T$_{1o}$	Si–O	4	2	1.614	4	–2(1)	0.0016(1)	Winter et al. (1977)
			T$_{1m}$	Si–O	4	2	1.615	4	–1(1)	0.0016(1)	Winter et al. (1977)
Feldspar	CaAl$_2$Si$_2$O$_8$	Anorthite		Ca–O	2	2	2.493	7	36(7)	0.0021(5)	Foit and Peacor (1973)
				Al–O	3	2	1.747	4	–8(6) }†	0.0008(2)	Foit and Peacor (1973)
				Si–O	4	2	1.614	4	13(6)	0.0006(2)	Foit and Peacor (1973)
Nephelline	(Na$_{5.6}$Ca$_{0.4}$K$_{1.3}$ □$_{0.3}$)Al$_8$Si$_8$O$_{32}$	Nephelline		Na–O	1	2	2.62	8	48(4)	0.0052(4)	Foreman and Peacor (1970)
Zeolite	Na$_2$Al$_2$Si$_3$O$_{10}$·2H$_2$O	Natrolite		Na–O	1	2	2.445	6	17(1)	0.0011(2)	Peacor (1973)
Cordierite	(Mg$_{1.9}$Fe$_{0.1}$)(Si$_5$Al$_4$)O$_{18}$ (M$_2$O$_{0.56}$)	Cordierite	M	(Mg, Fe)–O	2	2	2.108	6	13(1)	0.0015(1)	Hochella et al. (1979)
			T$_{1,1}$	Al–O	3	2	1.758	4	–3(3)	0.0012(1)	Hochella et al. (1979)
			T$_{1,6}$	Al–O	3	2	1.742	4	3(3)	0.0011(1)	Hochella et al. (1979)
			T$_{2,6}$	Si–O	4	1	1.626	4	4(3)	0.0010(1)	Hochella et al. (1979)
			T$_{2,1}$	Si–O	4	2	1.614	4	0(3)	0.0009(1)	Hochella et al. (1979)
			T$_{2,3}$	Si–O	4	2	1.617	4	1(1)	0.0009(1)	Hochella et al. (1979)
Mica	KMg$_3$AlSi$_3$O$_{10}$(OH)$_2$	Phlogopite	K	K–O	1	2	2.987	6	21(4)	0.0064(5)	Takeda and Morosin (1975)
			M$_1$	Mg–O	2	1.67	2.056	6	9(6)	0.0018(2)	Takeda and Morosin (1975)
			M$_2$	Mg–O	2	1.67	2.070	6	17(6)	0.0025(2)	Takeda and Morosin (1975)
			T	(Al$_{.25}$Si$_{.75}$)–O	3.75	2	1.651	4	–5(5)	0.0022(2)	Takeda and Morosin (1975)
Sillimanite	Al$_2$SiO$_5$	Sillimanite	Al1	Al–O	3	2	1.912	6	7(2)	0.0009(1)	Winter and Ghose (1979)
			Al2	Al–O	3	2	1.763	4	3(2)	0.0009(1)	Winter and Ghose (1979)
			Si	Si–O	4	2	1.627	4	1(1)	0.0008(1)	Winter and Ghose (1979)
Andalusite	Al$_2$SiO$_5$	Andalusite	Al1	Al–O	3	2	1.935	6	12(1)	0.0013(1)	Winter and Ghose (1979)
			Al2	Al–O	3	2	1.836	5	6(1)	0.0008(1)	Winter and Ghose (1979)
			Si	Si–O	4	2	1.631	4	0(1)	0.0007(1)	Winter and Ghose (1979)
Kyanite	Al$_2$SiO$_5$	Kyanite	Al1	Al–O	3	2	1.902	6	11(1)	0.0010(1)	Winter and Ghose (1979)
			Al2	Al–O	3	2	1.913	6	9(1)	0.0010(1)	Winter and Ghose (1979)
			Al3	Al–O	3	2	1.919	6	9(2)	0.0010(1)	Winter and Ghose (1979)
			Al4	Al–O	3	2	1.896	6	10(1)	0.0010(1)	Winter and Ghose (1979)
			Si1	Si–O	4	2	1.635	4	2(2)	0.0007(1)	Winter and Ghose (1979)
			Si2	Si–O	4	2	1.636	4	3(2)	0.0007(1)	Winter and Ghose (1979)

Table 6.2 continued

Bond	Structure	Formula	Site	z_c	z_a	d(Å)	n	$\bar{\alpha}^*_{1000} \times 10^6$ (K^{-1})	$\partial B/\partial T$ ($Å^2/K$)	Mineral Name	Reference
Li-O	Stuffed Quartz	$LiAlSiO_4$	Li1	1	2	1.964	4	19(8)	—	β-Eucryptite	Pillars and Peacor (1973)
Li-O			Li2	1	2	2.080	4	16(10)	—		Pillars and Peacor (1973)
Li-O			Li3	1	2	2.017	4	31(13)	0.0060(10)		Pillars and Peacor (1973)
Al-O			Al1	3	2	1.752	4	−2(4)	0.0016(2)		Pillars and Peacor (1973)
Al-O			Al2	3	2	1.713	4	4(6)	0.0018(2)		Pillars and Peacor (1973)
Si-O			Si1	4	2	1.640	4	−5(5)	0.0010(1)		Pillars and Peacor (1973)
Si-O			Si2	4	2	1.594	4	−8(8)	0.0009(1)		Pillars and Peacor (1973)
Mg-O	Amphibole	$Ca_2Mg_5Si_8O_{22}$ $(OH)_2$	M1	2	2	2.075	6	13(1)	0.0016(1)	Tremolite	Sueno et al. (1973)
Mg-O			M2	2	2	2.077	6	15(2)	0.0016(1)		Sueno et al. (1973)
Mg-O			M3	2	2	2.066	6	12(1)	0.0015(1)		Sueno et al. (1973)
Ca-O			M4	2	2	2.506	8	16(1)	0.0025(1)		Sueno et al. (1973)
Si-O			T1	4	2	1.620	4	0(1)	0.0014(1)		Sueno et al. (1973)
Si-O			T2	4	2	1.632	4	2(1)	0.0014(1)		Sueno et al. (1973)
Ba-O	Perovskite	$Ba_2Bi^{3+}Bi^{5+}O_6$	Ba	2	2	3.07	12	15(3)	0.0032(3)	—	Cox and Sleight (1979)
Bi-O			Bi1	3	2	2.28	6	10(6)	0.0022(4)		Cox and Sleight (1979)
Bi-O			Bi2	5	2	2.12	6	0(2)	0.0000(5)		Cox and Sleight (1979)

$$*\ \bar{\alpha}_{1000} = \frac{2}{d_0 + d_{1000}} \left(\frac{d_{1000} - d_0}{980} \right) \approx \alpha_{510},$$

where d_0 and d_{1000} are mean cation–anion bond distance at 20°C and 1000°C, respectively.

† These expansion coefficients are anomalous because of cation disordering at high temperature.

‡ Powder profile refinement.

§ These expansion coefficients are anomalous because of changes of structural topology at high temperature.

|| Several high-temperature structure studies involving phases that undergo transitions have not been included because bond distance variations are not continuous versus temperature. The structure studies include ($Ca_1Mg_4Fe_1$)-pigeonite (Brown et al., 1972); Bi_2UO_6 (Koster et al., 1975); $MgSiO_3$-pyroxenes (Smyth, 1969 and 1971); ($Mg_3Fe_{.7}$)SiO_3-clinohypersthene (Smyth and Burnham, 1972); and $Mg_3Si_8O_{22}(OH)_2$-cummingtonite (Sueno et al., 1972).

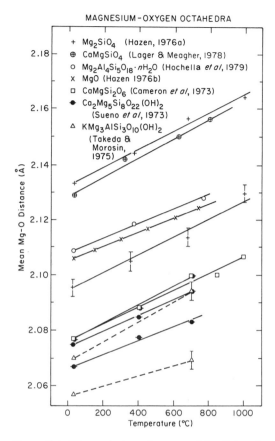

Figure 6-4. The variation of average Mg–O bond distances in seven different compounds with magnesium octahedra, plotted versus temperature

appears likely that these anomalies are due to disordering of Si and Al rather than changes in bonds of a fixed composition. The second positive expansion is $0.6(2) \times 10^{-5}\,^\circ C^{-1}$ in grossular garnet, which is still a small, and perhaps insignificant, expansion. The other three non-zero Si–O expansivities occur in the tetrahedral B-chains of three different two-chain pyroxenes. In each of these tetrahedral sites the O_3 (bridging oxygen) to silicon distance shows significant shortening with increasing temperature. This anomalous behaviour may be related to the changing topology with temperature of the O_3 oxygen. At room conditions in these pyroxenes the O_3 is one of seven oxygens coordinated to the large M2 site. With increasing temperature the M2–O_3 bond expands more than ten times faster than the average M2–O bond, indicating that O_3 is becoming dissociated from M2. This change in topology of the M2 site, from 7 to 6 coordinated, and the associated change in O_3 coordination from three to two cations, leads to the

Table 6-3. Mean linear expansion coefficients for cation polyhedra, based on data in Tables 6-1 and 6-2

Cation	Anion	n	$\dfrac{S^2 z_c z_a}{n}$	Number Obs.	Min $\bar{\alpha}_{1000}$	Max $\bar{\alpha}_{1000}$	Mean $\bar{\alpha}_{1000}$	$\dfrac{S^2 z_c z_a}{n} \times \bar{\alpha}_{1000} \times 10^6$
Re^{6+}	O^{2-}	6	1	1	–	–	1(2)	1(2)
Si^{4+}	O^{2+}	4	1	41	–8(8)	6(2)	0(2)	0(2)
Bi^{5+}	O^{2-}	6	5/6	1	–	–	0(2)	0(2)
Al^{3+}	O^{2-}	4	3/4	8	–8(6)	4(6)	1(2)	0.8(15)
Ti^{4+}	O^{2-}	6	2/3	5	6(2)	20(10)	8(2)	5.4(13)
Al^{3+}	O^{2-}	5	3/5	1	–	–	6(2)	3.6(11)
Zr^{4+}	O^{2-}	7	4/7	2	8(1)	8(1)	8(2)	4.6(11)
Hf^{4+}	O^{2-}	7	4/7	1	–	–	7(2)	4.0(12)
U^{4+}	O^{2-}	8	1/2	1	–	–	10(2)	5.0(10)
Th^{4+}	O^{2-}	8	1/2	3	8(1)	9(1)	9(2)	4.5(10)
Ce^{4+}	O^{2-}	8	1/2	1	–	–	13(2)	6.5(10)
V^{3+}	O^{2-}	6	1/2	1	–	–	13(2)	6.5(10)
Fe^{3+}	O^{2-}	6	1/2	2	8(2)	10(1)	9(2)	4.5(10)
Cr^{3+}	O^{2-}	6	1/2	2	7(1)	8(1)	8(2)	4.0(10)
Ti^{3+}	O^{2-}	6	1/2	1	–	–	8(2)	4.0(10)
Al^{3+}	O^{2-}	6	1/2	12	7(1)	12(1)	9(2)	4.5(10)
Bi^{3+}	O^{2-}	6	1/2	2	9(1)	10(6)	9(2)	4.5(10)
Be^{2+}	O^{2-}	4	1/2	2	8(1)	9(1)	9(2)	4.5(10)
Zn^{2+}	O^{2-}	4	1/2	2	7(1)	8(1)	7(2)	3.5(10)
Ni^{2+}	O^{2-}	6	1/3	5	13(1)	17(3)	14(1)	4.6(3)
Mg^{2+}	O^{2-}	6	1/3	20	9(6)	16(3)	14(1)	4.6(3)
Co^{2+}	O^{2-}	6	1/3	1	–	–	14(1)	4.6(3)
Fe^{2+}	O^{2-}	6	1/3	9	10(1)	16(4)	13(2)	4.3(7)
Mn^{2+}	O^{2-}	6	1/3	3	14(1)	15(1)	15(2)	5.0(7)
Cd^{2+}	O^{2-}	6	1/3	2	13(1)	13(1)	13(2)	4.3(7)
Ca^{2+}	O^{2-}	6	1/3	4	13(1)	19(1)	15(2)	5.0(7)
Ba^{2+}	O^{2-}	6	1/3	2	13(2)	18(2)	15(3)	5.0(10)
Sr^{2+}	O^{2-}	6	1/3	1	–	–	14(2)	4.6(7)
Ca^{2+}	O^{2-}	7	2/7	1	–	–	16(3)	4.6(8)
Ca^{2+}	O^{2-}	8	1/4	4	10(1)	16(1)	14(2)	3.5(5)
Mg^{2+}	O^{2-}	8	1/4	1	–	–	13(2)	3.3(5)
Li^+	O^{2-}	4	1/4	3	16(10)	31(13)	22(15)	6(4)
Ba^{2+}	O^{2-}	~9	2/9	1	–	–	15(3)	3.3(7)
Pb^{2+}	O^{2-}	12	1/6	1	–	–	23(8)	4(1)

Table 6-3   *continued*

Cation	Anion	n	$\dfrac{S^2 z_c z_a}{n}$	Number Obs.	Min $\bar{\alpha}_{1000}$	Max $\bar{\alpha}_{1000}$	Mean $\bar{\alpha}_{1000}$	$\dfrac{S^2 z_c z_a}{n} \times \bar{\alpha}_{1000} \times 10^6$
K^+	O^{2-}	6	1/6	1	–	–	21(4)	3.5(7)
Na^+	O^{2-}	6	1/6	1	–	–	17(3)	2.8(5)
Li^+	O^{2-}	6	1/6	1	–	–	20(3)	3.3(5)
Na^+	O^{2-}	7	1/7	1	–	–	35(2)	5.0(3)
Na^+	O^{2-}	8(1)	1/8	4	13(1)	48(4)	22(10)	3(1)
Na^+	O^{2-}	9(2)	1/9	2	17(1)	18(1)	18(3)	2(1)
Li^+	F^-	6	0.125	1	–	–	46	5.7
Na^+	Cl^-	6	0.125	2	–	–	51	6.4
K^+	Cl^-	6	0.125	2	–	–	46	5.7
K^+	Br^-	6	0.125	2	–	–	49	6.1
K^+	I^-	6	0.125	1	–	–	47	5.9
Na^+	F^-	6	0.125	1	–	–	45	5.6
Rb^+	Br^-	6	0.125	1	–	–	44	5.5
Ca^{2+}	F^-	8	0.19	2	–	–	21	3.9
Cs^+	Br^-	8	0.094	1	–	–	68	6.4
Mn^{2+}	S^{2-}	6	0.27	1	–	–	18	4.8
Pb^{2+}	S^{2-}	6	0.27	1	–	–	22	5.9
Pb^{2+}	Te^{2-}	6	0.27	1	–	–	20	5.3
Pb^{2+}	Se^{2-}	6	0.27	1	–	–	21	5.6
Zn^{2+}	S^{2-}	4	0.40	1	–	–	9	3.6
Zn^{2+}	S^{2-}	4	0.40	1	–	–	10	4.0
Zn^{2+}	Se^{2-}	4	0.40	1	–	–	9	3.6
Al^{3+}	As^{3-}	4	0.56	1	–	–	5	2.8
Ga^{3+}	As^{3-}	4	0.56	1	–	–	7	3.9
B^{3+}	N^{3-}	4	0.45	1	–	–	13	5.8
Ti^{4+}	N^{4-}	6	0.53	1	–	–	9	4.8
U^{4+}	N^{4-}	6	0.53	1	–	–	9	4.8
Nb^{4+}	C^{4-}	6	0.53	1	–	–	7	3.7
Ta^{4+}	C^{4-}	6	0.53	1	–	–	7	3.7
Ti^{4+}	C^{4-}	6	0.53	1	–	–	8	4.3
Zr^{4+}	C^{4-}	6	0.53	1	–	–	7	3.7
C^{4+}	C^{4-}	4	0.80	1	–	–	3.5	2.8

anomalous shortening of the SiB–O$_3$ bonds in these pyroxenes. The Si–O potential is changing, but the changes are due to external features not inherent in Si–O bonding.

An important generalization stemming from these observations is that the thermal expansion coefficient for each type of polyhedron is *independent of structural linkages* of the polyhedron, providing the site chemistry and nearest neighbour configuration of the structure do not change with temperature. Thus, for each type of cation–oxygen polyhedron there exists a value for an expansion coefficient that may be used to predict behaviour at high temperature. A list of observed polyhedral expansion coefficients is presented in Table 6-3.

A second generalization is evident from Table 6-3: all oxygen-based polyhedra with the same Pauling bond strength (cation valence, z_c, divided by coordination number, n) have the same $\bar{\alpha}_{1000}$. For example, octahedra of Ni, Mg, Co, Fe, Cd, Mn, Ca, Ba and Sr *all* have $\bar{\alpha}_{1000}$ of $14 \pm 1 \times 10^{-6}°C^{-1}$. This fact is remarkable given the range of atomic masses and size represented by these cations.

Hazen and Prewitt (1977) noted the two generalizations above, and proposed an empirical relationship relating the mean linear polyhedral thermal expansion coefficient, $\bar{\alpha}_{1000}$ and (z_c/n):

$$\bar{\alpha}_{1000} = 32.9 \ (0.75 - z_c/n) \times 10^{-6}°C^{-1} \tag{6-17}$$

This equation models the oxide and silicate data in Tables 6-1, 6-2 and 6-3 extremely well. However, the equation cannot be applied to anions other than oxygen, and the physical significance of the equation is obscure.

In an effort to derive a more general empirical equation for $\bar{\alpha}$ in terms of bond strength, thermal expansion data for halides, chalcogenides, nitrides and carbides were added to Tables 6-1 and 6-3. Again, for a given anion, all bonds with the same (z_c/n) have similar $\bar{\alpha}_{1000}$. It is also evident, as for oxides, that stronger bonds have smaller expansion coefficients.

A plot of $\bar{\sigma}_{1000}$ versus bond strength (Figure 6-5) reveals a simple inverse relationship between the two parameters:

$$\bar{\alpha}_{1000} \left(\frac{S^2 z_c z_a}{n} \right) = 4.0(4) \times 10^{-6}°C^{-1} \tag{6-18}$$

where S^2 is an ionicity factor (Hazen and Finger, 1979) *defined* to be 0.50 for all oxides and silicates, and *observed* to be 0.75 for all halides, 0.40 for chalcogenides, 0.25 for phosphides and arsenides, and 0.20 for nitrides and carbides. (Bond compression data are used to derive S^2 values, as discussed in Chapter 7, section II-B.)

Equation (6-18) is physically reasonable. If bond strength is zero between two atoms, as in the case of an inert gas, then thermal expansion is infinite. If bond strength is very large, then thermal expansion approaches zero. In practice equation (6-18) may be used to predict the linear thermal expansion coefficients of most polyhedra to within 20 per cent. The formula does not work well for very large alkali sites, for which coordination number may not be well defined. The

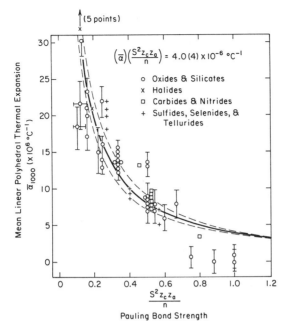

Figure 6-5. The inverse relationship between polyhedral thermal expansion and Pauling bond strength

formula is also inadequate for bond strengths greater than 0.75, which are observed to have expansion coefficients less than those predicted by equation (6-18). Yet another limitation of this bond strength–thermal expansion relationship is the lack of information on thermal corrections to bond distances. Actual expansion coefficients must be somewhat larger than those recorded in Tables 6-1 to 6-3 for uncorrected bond distances.

C. Relationships between polyhedral and bulk thermal expansion

Equation (6-18) allows the prediction of polyhedral thermal expansion, given Pauling bond strength. Bulk crystal thermal expansion, as well as bulk modulus, does not generally result entirely from changes in cation–anion distances, however; changes in angles between polyhedra must also be considered.

Two cation polyhedra may be linked by a shared face, a shared edge, a shared corner or Van der Waal's forces. The type and distribution of these polyhedral linkages are the most important factors in determining the bulk thermal expansion of a compound. The effects of different linkages on expansion of two-dimensional analogues is illustrated in Figure 6-6 (a–d). The degree to which mineral bulk expansion differs from polyhedral bulk expansion depends on the rigidity of polyhedral linkages.

138

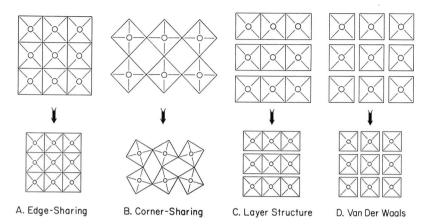

| A. Edge-Sharing | B. Corner-Sharing | C. Layer Structure | D. Van Der Waals |

Figure 6-6. Effects of polyhedral linkage on bulk expansion or compression. The square (a two-dimensional analogue of the polyhedron) undergoes a constant area change in each of the four examples. The total area change is not constant owing to the differences in linkages of the squares. Similarly, volume changes of a compound with temperature or pressure will depend on polyhedral linkages as well as volume changes of the polyhedra themselves

The most rigid polyhedral linkage is one in which polyhedra share faces or edges in three dimensions (Figure 6-6a is a two-dimensional analogue). If each shared edge between polyhedra is represented in space as a line segment, then all such line segments may form a continuous three-dimensional array. In these fully edge-linked structures (including rock salt, corundum, spinel and garnet) any change in molar volume must be accompanied by a change in metal–oxygen distance because of rigid polyhedral linkages. Thermal expansion of these compounds, consequently, is small because it is similar in magnitude to the thermal expansion of metal–oxygen polyhedra. In rock salt- and corundum-type compounds, for example, the thermal expansivities of the octahedra, the only polyhedra, and the bulk compound are identical. In silicate spinels and garnets, having both silicate tetrahedra and larger divalent cation polyhedra, the mineral expansivities are intermediate between the polyhedral expansivities of these Si^{4+} and R^{2+} sites.

In contrast to fully edge-linked structures, some materials such as α-quartz, feldspar and zeolites have primarily corner-linked polyhedra (Figure 6-6b). In these framework structures volume changes may be affected by changes in angles between tetrahedra, without altering T–O distances. Framework silicates, consequently, have relatively large thermal expansion, even though individual tetrahedra may undergo no volume change with temperature. The tilting of polyhedra in expansion or compression of corner-linked materials may be treated as primarily metal–oxygen–metal bond bending, as opposed to metal–oxygen bond expansion or compression.

In most structures, including layer, chain and orthosilicates, all polyhedra share edges with some adjacent polyhedra, and link corners with others (Figure 6-6c). A continuous three-dimensional edge linkage does not obtain. In these materials expansion is due to a combination of polyhedral (metal–oxygen) expansion and bond bending, and the net expansion is greater than that of component polyhedra. The significant differences in polyhedral linkages between the olivine and spinel forms of Mg_2SiO_4, for example, lead to the greater expansion of the former.

Effects of linkage rigidity on thermal expansion are dramatically illustrated by layer minerals. In materials such as CdI_2, $Mg(OH)_2$, or layer silicates, weak interlayer bonds expand many times more than expansion within the layers. Phlogopite mica—$KMg_3AlSi_3O_{10}$ $(OH)_2$, for example, has fully linked edge-shared octahedra of magnesium in two dimensions within layers, but relatively weak K–O bonds between layers. The intralayer thermal expansion is similar to that of MgO, which is fully linked in three dimensions. The interlayer linear expansion of phlogopite, however, is several times greater, owing to the large expansivity of K–O bonds relative to Mg–O.

IV. THERMAL PARAMETER VARIATION WITH TEMPERATURE

Isotropic temperature factors are routinely reported in crystallographic studies as a measure of the time-averaged displacement of electron density from the mean centric position. As noted previously, this parameter may reflect either positional disorder or thermal vibrations. If vibrations are the cause of increased temperature factors, then the variation of these factors with temperature provides a direct measure of the increased thermal energy of an atom. In most compounds with fully ordered sites, a plot of isotropic temperature factor, B, versus temperature (in K) passes near the origin, indicating zero thermal vibration at absolute zero. Typical B versus T plots are illustrated in Figure 6-7.

It is useful to compare the slopes of these lines, $\partial B/\partial T$, in several different types of sites. Table 6-2 lists these slopes for several dozen cations. The first important generalization from these data, is that $\partial B/\partial T$ for a given polyhedron, like the expansion coefficient, is relatively constant. Figure 6-8, for example, illustrates the variation of B for nine magnesium cations in octahedral coordination versus temperature. The slopes are remarkably similar, having an average value of 0.0015 $A^2 K^{-1}$, with a range of ± 0.0002 $A^2 K^{-1}$. Similarly values of $\partial B/\partial T$ for octahedra of Ni, Fe and Ca cluster about 0.0012, 0.0015 and 0.0022 $A^2 K^{-1}$, respectively. Observed values of $\partial B/\partial T$ for several cations in oxygen polyhedra are listed in Table 6-4. Cations are listed in Table 6-4 only if values of $\partial B/\partial T$ are available from at least two different structures. Cations from sites of changing occupancy or from structures of changing topology, are not included.

The uniformity of $\partial B/\partial T$ for like polyhedra in different structures was predicted by Ohashi and Finger (1973) on the basis of their analysis of thermal vibrations. The isotropic temperature factor, B, is related to mean-square displacement (m.s.d.) $\langle r^2 \rangle$, by:

$$B = 8\pi^2 \langle r^2 \rangle \qquad (6-19)$$

140

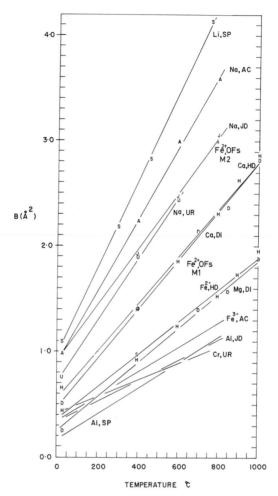

Figure 6-7. Variation with increasing temperature of the equivalent isotropic temperature factors of octahedral and 8-coordinated cations in six chain silicates. Abbreviations are $SP = LiAlSi_2O_6$ (spodumene), $AC = NaFeSi_2O_6$ (acmite), $JD = NaAlSi_2O_6$ (jadeite), $HD = CaFeSi_2O_6$ (hedenbergite), $DI = CaMgSi_2O_6$ (diopside), and $UR = NaCrSi_2O_6$ (ureyite). (Reproduced from Cameron *et al.*, 1973, by permission of the Mineralogical Society of America)

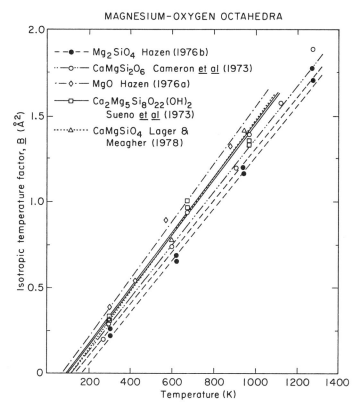

Figure 6-8. Isotropic temperature factor, B, versus temperature for magnesium in octahedral coordination

and the m.s.d. is, in turn, a function of the bond force constant, f:

$$\langle r^2 \rangle = kT/f \qquad (6\text{-}20)$$

where k is the Boltzmann constant and T is absolute temperature. Thus:

$$B = 8\pi^2 kT/f \qquad (6\text{-}21)$$

If f is independent of T, then:

$$\partial B/\partial T = 8\pi^2 k/f \qquad (6\text{-}22)$$

For a given type of cation polyhedron f, and thus $\partial B/\partial T$, is similar in all structures. The bond force constant is roughly proportional to cation valence and inversely proportional to cation–anion distance. Therefore, as observed in Table 6-4, the variation of B with T is greater for monovalent cations than for divalent cations, which in turn show greater variation than 3+ and 4+ cations. For a polyhedron of a given cation charge, such as the several R^{2+} octahedra, $\partial B/\partial T$ increases with increasing cation–anion bond distance as predicted.

Table 6-4. Variation of isotropic temperature factors versus temperature for cations in cation–anion polyhedra

Cation	n	d_{M-O}	$(\partial B/\partial T)$	
Si^{4+} in framework silicates	4	1.61	0.0014	⎫ Average
Si^{4+} in chain silicates	4	1.63	0.0011	⎬ 0.0011
Si^{4+} in orthosilicates	4	1.65	0.0009	⎭
Ti^{4+}	6	1.96	0.0014	
Al^{3+} in framework	4	1.72	0.0015	⎱ Average
Al^{3+} in other	4	1.74	0.0010	⎰ 0.0012
Al^{3+}	6	1.91	0.0010	
Fe^{3+}	6	2.02	0.0012	
Ni^{2+}	6	2.09	0.0012	
Mg^{2+}	6	2.11	0.0015	
Fe^{2+}	6	2.17	0.0015	
Mn^{2+}	6	2.21	0.0019	
Ca^{2+}	6	2.36	0.0022	
Ca^{2+}	8	2.50	0.0025	
Na^+	8	2.62	0.0030	
Na^+	9	2.81	0.0045	

A single value of $\partial B/\partial T$ may not be adequate to describe silicon tetrahedra. An overall average value of $0.0011\ A^2\ K^{-1}$ is observed, but $\partial B/\partial T$ in framework, chain and orthosilicates appears to be different. Data in Table 6-2 yield average values of 0.0014, 0.0011 and $0.0009\ A^2\ K^{-1}$, respectively, for these three distinct types of tetrahedral linkages. The principal differences between these silicates is the number of bridging oxygens (i.e. the number of oxygens that link two tetrahedra). In framework silicates all four oxygens in each tetrahedra are bridging, whereas in chain silicates only two oxygens bridge tetrahedra, and in orthosilicates there are no bridging oxygens. Thus $\partial B/\partial T$ appears to be a function of polyhedral linkages as well as of cation valence and bond distance.

V. CONCLUSIONS

Continuous variation of atomic positions with temperature in ionically bonded compounds are regular, systematic and predictable. It has been demonstrated from thermal expansion data on bonds in both simple and complex structures that the mean linear expansion coefficient of a polyhedron is approximately

$$\alpha_{1000} = 4.0\ (4) \left[\frac{n}{S^2 z_c z_a} \right] \times 10^{-6\circ}C^{-1}$$

independent of polyhedral topology, provided site compositions and topologies do not change on heating. Polyhedral thermal expansion is inversely proportional to the Pauling bond strength of the cation–anion bonds within the polyhedron.

An important conclusion of this chapter is that the cation polyhedron, a basic building block of most ionic structures, has certain physical properties that are independent of structural linkage. These polyhedral properties include volume, shape, thermal expansion coefficient, and thermal vibration parameters and their temperature variation. The constancy of these parameters for a given polyhedron in different structures indicates the great control of nearest-neighbour interactions in determining the atomic-scale properties of ionically bonded compounds.

In view of the importance of the polyhedral unit in characterizing the properties of an ionic solid, changes in structure with temperature may be modelled as the variation in the *ratio* of polyhedral sizes, due to differing polyhedral expansion coefficients. Thus, for example, in a silicate with magnesium octahedra, the ratio of octahedral to tetrahedral size increases with temperature. The utility of this approach to the temperature variation of crystal structure will be explored in Chapters Nine and Ten.

REFERENCES

Ackermann, R. J., and C. A. Sorrell (1974) Thermal expansion and the high-low transformation in quartz. I. High-temperature x-ray studies. *J. Appl. Cryst.* **7**, 461–467.

Austin, J. B. (1931) Thermal expansion of some refractory oxides. *J. Am. Ceram. Soc.* **14**, 795–810.

Beals, R. J., and R. L. Cook (1957) Directional dilatation of crystal lattices at elevated temperatures. *J. Am. Ceram. Soc.* **40**, 279–284.

Bobrovskii, A. B., G. N. Kartmazov and V. A. Finkel (1973) Crystal structure of nickel monoxide at high temperature. *Izv. Akad. Nauk SSSR Neorg. Mater.* **9**, 1075–1076.

Brooksbank, D., and K. W. Andrews (1968) Thermal expansion of some inclusions found in steels and relation to tessellated stresses. *J. Iron and Steel Inst.* June, 595–599.

Brown, G. E., and C. T. Prewitt (1973) High-temperature crystal chemistry of hortonolite. *Am. Mineral.* **58**, 577–587.

Brown, G. E., C. T. Prewitt, J. J. Papike and S. Sueno (1972) A comparison of the structures of low and high pigeonite. *J. Geophys. Res.* **77**, 5778–5789.

Busing, W. R., and H. A. Levy (1964) The effect of thermal motion on the estimation of bond lengths from diffraction measurements. *Acta Crystallogr.* **17**, 142–146.

Cameron, M., S. Sueno, C. T. Prewitt and J. J. Papike (1973) High-temperature crystal chemistry of acmite, diopside, hedenbergite, jadeite, spodumene and ureyite. *Am. Mineral.* **58**, 594–618.

Campbell, W. J., and C. F. Grain (1961) Thermal expansion of α-alumina. *U.S. Bureau Mines Rept. Investigation* 5757, 16 pp.

Carter, R. E. (1959) Thermal expansion of $MgFe_2O_4$, FeO, and $MgO \cdot 2FeO$. *J. Am. Ceram. Soc.* **42**, 324–327.

Chang, T.-S., and P. Trucano (1978) Lattice parameter and thermal expansion of ReO_3 between 291 and 464 K. *J. Appl. Cryst.* **11**, 286–288.

Cox, D. E., and A. W. Sleight (1979) Mixed-valent $Ba_2Bi^{3+}Bi^{5+}O_6$: structure and properties versus temperature. *Acta Crystallogr.* **B35**, 1–10.

Das, C. D., H. V. Keer and R. V. G. Rao (1963) Lattice energy and other properties of some ionic crystals. *Z. Physik. Chem.* **224**, 377–383.

Deganello, S. (1978) Thermal expansion from 25° to 500°C of a few ionic radii. *Z. Kristallogr.* **147**, 217–227.

144

Eisenstein, A. J. (1946) A study of oxide cathods by x-ray diffraction methods. *J. Appl. Phys.* **17**, 434–443.

Ettenberg, M., and R. J. Paff (1970) Thermal expansion of AlAs. *J. Appl. Phys.* **41**, 3926–3927.

Feder, R., and T. Light (1968) Precision thermal expansion measurements of semi-insulating GaAs. *J. Appl. Phys.* **39**, 4870–4871.

Finger, L. W., R. M. Hazen and Y. Yagi (1979) Crystal structures and electron densities of nickel and iron silicate spinels at elevated temperature or pressure. *Am. Mineral.* **64**, 1002–1009.

Foit, F. F., and D. R. Peacor (1973) The anorthite crystal structure at 410 and 830°C. *Am. Mineral.* **58**, 665–675.

Foreman, N., and D. R. Peacor (1970) Refinement of the nepheline structure at several temperatures. *Z. Kristallogr.* **132**, 45–70.

Glazer, A. M., and S. A. Mabud (1978) Powder profile refinement of lead zirconate titanate at several temperatures. II. Pure PbTiO$_3$. *Acta Crystallogr.* **B34**, 1065–1070.

Grain, C. F., and W. J. Campbell (1962) Thermal expansion and phase inversions of six refractory oxides. *U.S. Bureau Mines Rept. Investigation* 5982, 21 pp.

Hanneman, R. E., and H. C. Gatos (1965) The relation between compressibility and thermal expansion coefficients in cubic metals and alloys. *J. Appl. Phys.* **36**, 1794–1796.

Hazen, R. M. (1975) *Effects of Temperature and Pressure on the Crystal Physics of Olivine*, Ph.D. Thesis, Department of Geological Sciences, Harvard University.

Hazen, R. M. (1976a) Effects of temperature and pressure on the cell dimension and x-ray temperature factors of periclase. *Am. Mineral.* **61**, 266–271.

Hazen, R. M. (1976b) Effects of temperature and pressure on the crystal structure of forsterite. *Am. Mineral.* **61**, 1280–1293.

Hazen, R. M., and L. W. Finger (1979) Bulk modulus-volume relationship for cation-anion polyhedra. *J. Geophys. Res.* **84**, 6723–6728.

Hazen, R. M., and C. T. Prewitt (1977) Effects of temperature and pressure on interatomic distances in oxygen-based minerals. *Am. Mineral.* **62**, 309–315.

Hochella, M. F., G. E. Brown, F. K. Ross and G. V. Gibbs (1979) High-temperature crystal chemistry of hydrous Mg- and Fe-cordierites. *Am. Mineral.* **64**, 337–351.

Horn, M., C. F. Schwerdtfeger and E. P. Meagher (1972) Refinement of the structure of anatase at several temperatures. *Z. Kristallogr.* **136**, 273–281.

Houston, B., R. E. Strakna and H. S. Belson (1968) Elastic constants, thermal expansion, and Debye temperature of lead telluride. *J. Appl. Phys.* **39**, 3913–3916.

Jun, C. K. (1970) Thermal expansion of NbC, HfC, and TaC at high temperatures. *J. Appl. Phys.* **41**, 5081–5082.

Kempter, C. P., and R. O. Elliott (1959) Thermal expansion of UN, UO$_2$, UO$_2 \cdot$ ThO$_2$, and ThO$_2$. *J. Chem. Phys.* **30**, 1524–1526.

Kittel, C. (1971) *Introduction to Solid State Physics*, John Wiley and Sons, NY.

Koster, A. S., J. P. P. Renaud and G. D. Rieck (1975) The crystal structures at 20 and 1000°C of bismuth uranate, Bi$_2$UO$_6$. *Acta Crystallogr.* **B31**, 127–131.

Kumar, S. (1959) Thermal expansion of simple ionic crystals. *Proc. Natl. Inst. Sci. India* **A25**, 364–372.

Lager, G. A., and E. P. Meagher (1978) High-temperature structural study of six olivines. *Am. Mineral.* **63**, 365–377.

Larionov, A. L., and B. Z. Malkin (1975) Thermal expansion of calcium fluoride. *Phys. Stat. Sol.* **B60**, K103–K105.

Lonsdale, K. (1959) Experimental studies of atomic vibrations in crystals and of their relations to thermal expansion. *Z. Kristallogr.* **112**, 188–212.

Meagher, E. P. (1975) The crystal structures of pyrope and grossularite at elevated temperatures. *Am. Mineral.* **60**, 218–228.

Meagher, E. P., and G. A. Lager (1979) Polyhedral thermal expansion in the TiO_2 polymorphs: refinement of the crystal structures of rutile and brookite at high temperature. *Canadian Mineral.* **17**, 77–85.

Megaw, H. D. (1938) The thermal expansion of crystals in relation to their structure. *Z. Kristallogr.* **A100**, 58–76.

Min'ko, N. I. (1972) Change in interion distances in oxides in the 298°–1773°K range. *Zh. Fiz. Khim.* **46**, 312–315.

Nielsen, T. H., and M. H. Leipold (1965) Thermal expansion of nickel oxide. *J. Am. Ceram. Soc.* **48**, 164.

Ohashi, Y., and C. W. Burnham (1972) Electrostatic and repulsive energies of the M1 and M2 cation sites in pyroxenes. *J. Geophys. Res.* **77**, 5761–5766.

Ohashi, Y., and L. W. Finger (1973) Thermal vibration ellipsoids and equipotential surfaces at the cation sites in olivine and clinopyroxenes. *Carnegie Inst. Washington Year Book* **72**, 547–551.

Pathak, P. D., and N. G. Vasavada (1970) Thermal expansion of NaCl, KCl, and CsBr by x-ray diffraction and the law of corresponding states. *Acta Crystallogr.* **A26**, 655–658.

Pathak, P. D., J. M. Trivedi and N. G. Vasavada (1973) Thermal expansion of NaF, KBr, and RbBr and temperature variation of the frequency spectrum of NaF. *Acta Crystallogr.* **A29**, 477–479.

Pauling, L. (1960) *The Nature of the Chemical Bond*, Cornell University Press, Ithaca, NY.

Peacor, D. R. (1973) High-temperature single-crystal x-ray study of natrolite. *Am. Mineral.* **58**, 676–680.

Pease, R. S. (1952) X-ray study of boron nitride. *Acta Crystallogr.* **5**, 356–361.

Petukhov, V. A. (1973) Single crystal aluminum oxide as a standard substance in dilatometry. *Teplofiz. Vys. Temp.* **11**, 1083–1087.

Pillars, W. W., and D. R. Peacor (1973) The crystal structure of β-Eucryptite as a function of temperature. *Am. Mineral.* **58**, 681–690.

Prewitt, C. T., S. Sueno and J. J. Papike (1976) The crystal structures of high albite and monalbite at high temperatures. *Am. Mineral.* **61**, 1213–1225.

Rice, C. E., and W. R. Robinson (1977) High-temperature crystal chemistry of Ti_2O_3: Structural changes accompanying the semiconductor-metal transition. *Acta Crystallogr.* **B33**, 1342–1348.

Rigby, G. R., G. H. B. Lovell and A. T. Green (1946) Reversible thermal expansion and other properties of some magnesian ferrous silicates. *Trans. British Ceram. Soc.* **45**, 237–250.

Robinson, W. R. (1975) High-temperature crystal chemistry of V_2O_3 and 1 per cent chromium-doped V_2O_3. *Acta Crystallogr.* **B31**, 1153–1160.

Sharma, S. S. (1950) Thermal expansion of crystals. *Proc. Indian Acad. Sci.* **32**, 268–274.

Singh, H. P., and B. Dayal (1969) Lattice parameters of cadmium oxide at elevated temperatures. *Solid State Commun.* **7**, 725–726.

Skinner, B. J. (1957) The thermal expansions of thoria, periclase and diamond. *Am. Mineral.* **42**, 39–55.

Skinner, B. J. (1962) Thermal expansion of ten minerals. *U.S. Geological Survey Professional Paper* **450D**, 109–112.

Skinner, B. J. (1966) Thermal expansion. In *Handbook of Physical Constants*, S. P. Clark (ed), *Geological Soc. Am. Memoir* **97**, 78–96.

Smyth, J. R. (1969) Orthopyroxene-high-low clinopyroxene inversions. *Earth Planet. Sci. Lett.* **6**, 406–407.

Smyth, J. R. (1971) Protoenstatite: a crystal-structure refinement at 1100°C. *Z. Kristallogr.* **134**, 262–274.

Smyth, J. R. (1973) An orthopyroxene structure up to 850°C. *Am. Mineral.* **58**, 636–848.

146

Smyth, J. R. (1974) The high-temperature crystal chemistry of clinohypersthene. *Am. Mineral.* **59**, 1069–1082.

Smyth, J. R. (1975) High temperature crystal chemistry of fayalite. *Am. Mineral.* **60**, 1092–1097.

Smyth, J. R., and C. W. Burnham (1972) The crystal structures of high and low clinohypersthene. *Earth Planet. Sci. Lett.* **14**, 183–189.

Smyth, J. R., and R. M. Hazen (1973) The crystal structures of forsterite and hortonolite at several temperatures up to 900°C. *Am. Mineral.* **58**, 588–593.

Stecura, S., and W. J. Campbell (1962) Thermal expansion and phase inversion of rare-earth oxides. *U.S. Bureau Mines Rept. Investigation* 5847, 47 pp.

Sueno, S., M. Cameron, J. J. Papike and C. T. Prewitt (1973) The high temperature crystal chemistry of tremolite. *Am. Mineral.* **58**, 649–664.

Sueno, S., M. Cameron and C. T. Prewitt (1976) Orthoferrosilite: high temperature crystal chemistry. *Am. Mineral.* **61**, 38–53.

Sueno, S., J. J. Papike, C. T. Prewitt and G. E. Brown (1972) Crystal structure of cummingtonite. *J. Geophys. Res.* **77**, 5767–5777.

Suzuki, I. (1975) Thermal expansion of periclase and olivine, and their anharmonic properties. *J. Phys. Earth* **23**, 145–159.

Takeda, H., and B. Morosin (1975) Comparison of observed and predicted structural parameters of mica at high temperature. *Acta Crystallogr.* **B31**, 2444–2452.

Taylor, M., and G. E. Brown (1976) High-temperature structural study of the $P2_1/a$-$A2/a$ phase transition in synthetic titanite, $CaTiSiO_5$. *Am. Mineral.* **61**, 435–447.

Valeev, K. S., and V. B. Kvaskov (1973) Thermal expansion of bismuth, cadmium and zinc oxides. *Izv. Akad. Nauk. SSSR, Neog. Mater.* **9**, 714–715.

Van Uitert, L. G., H. M. O'Bryan, H. J. Guggenheim, R. L. Barns and G. Zydzik (1977a) Correlation of the thermal expansion coefficients of rare earth and transition metal oxides and fluorides. *Mater. Res. Bull.* **12**, 307–314.

Van Uitert, L. G., H. M. O'Bryan, M. E. Lines, H. J. Guggenheim and G. Zydzik (1977b) Thermal expansion—an empirical correlation. *Mater. Res. Bull.* **12**, 261–268.

Wechsler, B. A. (1977) Cation distribution and high-temperature crystal chemistry of armalcolite. *Am. Mineral.* **62**, 913–920.

White, G. K. (1973) Thermal expansion of reference materials: copper, silica and silicon. *J. Phys.* **D6**, 2070–2076.

Winter, J. K., and S. Ghose (1979) Thermal expansion and high-temperature crystal chemistry of the Al_2SiO_5 polymorphs. *Am. Mineral.* **64**, 573–586.

Winter, J. K., S. Ghose and F. P. Okamura (1977) A high-temperature study of the thermal expansion and the anisotropy of the sodium atom in low albite. *Am. Mineral.* **62**, 921–931.

Winter, J. K., F. P. Okamura and S. Ghose (1979) A high-temperature structural study of high albite, monalbite, and the analbite–monalbite phase transition. *Am. Mineral.* **64**, 409–423.

Zollweg, R. J. (1955) X-ray lattice constant of barium oxide. *Phys. Rev.* **100**, 671–673.

Chapter 7

Structural Variations with Pressure

CONTENTS

I   Effects of Pressure on the Ionic Bond 147
   A   Definition of Compressibility 147
   B   Calculated Compressibility of the Ionic Bond 148
II Dimensional Variations with Pressure 149
   A   Review of Previous Studies 149
      (1)   Compressibilities of fixed-structure compounds 149
      (2)   High-pressure crystal structures 149
   B   Empirical Bond Distance–Pressure Relationships 151
   C   Relationships Between Polyhedral and Crystal Bulk Modulus 159
      (1)   Calculation of crystal bulk moduli 159
      (2)   Pressure derivatives of bulk moduli 160
III Other Structural Variations with Pressure 160
   A   Polyhedral Distortions 160
   B   Variation of Temperature Factors with Pressure 161
IV Conclusions 161

I.   EFFECTS OF PRESSURE ON THE IONIC BOND

A.   Definition of compressibility

Compressibility, or the coefficient of pressure expansion, is defined in a way analogous to the coefficient of thermal expansion [equations (6-11), (6-12) and (6-13)]:

$$\text{Linear: } \beta_l = -\frac{1}{d}\left(\frac{\partial d}{\partial P}\right)_{T,X} \tag{7-1}$$

$$\text{Volume: } \beta_v = -\frac{1}{V}\left(\frac{\partial V}{\partial P}\right)_{T,X} \tag{7-2}$$

$$\text{Mean: } \bar{\beta}_v = \frac{-2}{(V_1 + V_2)} \left(\frac{V_2 - V_1}{P_2 - P_1} \right) \tag{7-3}$$

The compressibility of any linear or volume element of a crystal may thus be determined.

An important parameter that relates the change of volume with pressure is the bulk modulus, K in units of pressure, which is simply the inverse of mean compressibility:

$$\text{Bulk Modulus: } K = \bar{\beta}_v^{-1} \tag{7-4}$$

Bulk modulus is closely related to Young's modulus and the shear modulus and is thus commonly cited in physics and geophysics literature. Compressibility is more widely used in the chemical thermodynamics literature, however, because of its close functional relationship to thermal expansivity.

B. Calculated compressibility of the ionic bond

If a pressure (force per unit area, P) acts on a cross-sectional area approximately equal to the square of the interionic distance (d^2) then the net force on the bond is $F_p = Pd^2$. At equilibrium distance the sum of the bonding forces is zero:

$$\frac{\partial U_{Total}}{\partial d_{1-2}} + F_p = 0 \tag{7-5}$$

Combining equation (7-5) with the expression for site potential energy [equation (6-9)]:

$$\frac{Ae^2}{d^2} - \frac{nB}{d^{(n+1)}} + Pd^2 = 0 \tag{7-6}$$

or,

$$P = \frac{nB}{d_{1-2}^{(n+3)}} - \frac{Ae^2}{d_{1-2}^4} \tag{7-7}$$

The fractional change of interionic distance with pressure is:

$$-\frac{1}{d_{1-2}(\partial P/\partial d_{1-2})} = \frac{1}{\left[\dfrac{(n+3)nB}{d_{1-2}^{(n+3)}} - \dfrac{4Ae^2}{d_{1-2}^4} \right]} \tag{7-8}$$

which, combined with equation (7-7), gives:

$$-\frac{1}{d} \left(\frac{\partial d}{\partial P} \right) = \frac{1/(n+3)}{P + [(n-1)/(n+3)] \left(\dfrac{Ae^2}{d_{1-2}^4} \right)} \tag{7-9}$$

Thus, if values of bonding parameters n, B, d and A are known, then bond compressibility may be calculated. In practice equation (7-9) is adequate for the prediction of relative, but not absolute, values of bond compressibility (Hazen, 1975). The two-term bond potential model of equation (6-9) is too simple for most applications. Das *et al.* (1963), who used a four-term potential model, achieved somewhat better results in their compressibility calculations for NaCl-type oxides and halides. More sophisticated bonding models, such as the self-consistent symmetrized augmented plane wave method (Bukowinski, 1980) and the modified electron-gas method (Tossell, 1980), have proved more successful in predicting compression of ionic bonds in simple compounds.

II. DIMENSIONAL VARIATIONS WITH PRESSURE

A. Review of previous studies

(1) Compressibilities of fixed-structure compounds

Bond compressibilities may be derived from two types of experimental studies, as noted for thermal expansivity in the previous chapter. Simple compounds with fixed structure types, such as NaCl, CsCl, CaF_2 and ZnS, have bond compressibilities that are equal to the linear unit-cell compressibility. Studies of the bulk elastic moduli or compression of these compounds, therefore, also yield information on the variation of bond distance with pressure. Birch (1966), Simmons and Wang (1971) and Hazen and Finger (1979a) have compiled data, which include compressibilities of simple fixed compounds.

(2) High-pressure crystal structures

Complete three-dimensional crystal structure refinement is the only method for obtaining bond compressibilities in most materials. Only about 30 compounds have been studied in this way (Table 7-1), but the number is steadily growing. At present the great majority of high-pressure structure refinements have been performed on oxides and silicates because of the interest of geoscientists in the high-pressure behaviour of minerals.

An important result of structure studies at high pressure is that most compounds do not compress uniformly. Rather than simply scaling to smaller dimensions with increasing pressure, different bonds and angles in most structures change by different amounts. In magnesium silicates such as forsterite (Mg_2SiO_4) and enstatite ($MgSiO_3$), for example, the magnesium octahedra are significantly more compressible than the silica tetrahedra (Hazen, 1976a; Ralph and Ghose, 1980). Similarly, in the mica-type layer silicates, which have large alkali sites, octahedral divalent cation sites, and tetrahedral ($AlSi_3$) sites, the alkali sites are twice as compressible as the octahedral sites, which are in turn more compressible than the tetrahedra (Hazen and Finger, 1978a). Data on the *differential* compression of bonds in these nonfixed structures are essential in developing empirical relationships for the prediction of crystal structure variation with pressure.

Table 7-1. Complete three-dimensional, high-pressure (> 10 kbar), single-crystal, structure refinements*

Mineral Name	Formula	Structure Type	Reference
Antimony	Sb	Arsenic	Schiferl (1977)
Selenium	Se	Selenium	Keller et al. (1977)
Tellurium	Te	Selenium	Keller et al. (1977)
Periclase	MgO	Rock Salt	Hazen (1976b)
Halite	NaCl	Rock Salt	Finger and King (1978)
Troilite	FeS	Nickel Arsenide	King (1978)
Ruby	Al_2O_3	Corundum	Finger and Hazen (1978)
Ruby	Al_2O_3	Corundum	D'Amour et al. (1978)
Hematite	Fe_2O_3	Corundum	Finger and Hazen (1980)
Eskolaite	Cr_2O_3	Corundum	Finger and Hazen (1980)
Karelianite	V_2O_3	Corundum	Finger and Hazen (1980)
Quartz	SiO_2	Quartz	D'Amour et al. (1979)
Quartz	SiO_2	Quartz	Levien et al. (1980)
Coesite	SiO_2	Coesite	Levien and Prewitt (1981b)
Rutile	TiO_2	Rutile	Hazen and Finger (1981)
Cassiterite	SnO_2	Rutile	Hazen and Finger (1981)
–	RuO_2	Rutile	Hazen and Finger (1981)
–	GeO_2	Rutile	Hazen and Finger (1981)
–	MnF_2	Rutile	Hazen et al. (1978)
–	$Cs_2Au_2Cl_6$	Caesium Gold Chloride	Denner et al. (1979)
Calcite	$CaCO_3$	Calcite	Merrill and Bassett (1975)
Forsterite	Mg_2SiO_4	Olivine	Hazen (1976a)
Forsterite	Mg_2SiO_4	Olivine	Hazen and Finger (1980)
Fayalite	Fe_2SiO_4	Olivine	Hazen (1977)
–	Fe_2SiO_4	Spinel	Finger et al. (1977, 1979)
–	Ni_2SiO_4	Spinel	Finger et al. (1977, 1979)
Zircon	$ZrSiO_4$	Zircon	Hazen and Finger (1979c)

Table 7-1   *continued*

Mineral Name	Formula	Structure Type	Reference
Pyrope	$Mg_3Al_2Si_3O_{12}$	Garnet	Hazen and Finger (1978c)
Grossularite	$Ca_3Al_2Si_3O_{12}$	Garnet	Hazen and Finger (1978c)
Enstatite	$MgSiO_3$	Pyroxene	Ralph and Ghose (1980)
Diopside	$CaMgSi_2O_6$	Pyroxene	Levien and Prewitt (1981a)
Fassaite	$(Ca, Mg, Fe, Ti, Al)_2(Si, Al)_2O_6$	Pyroxene	Hazen and Finger (1977)
Gillespite	$BaFeSi_4O_{10}$	Gillespite	Hazen and Burnham (1974, 1975)
Phlogopite	$KMg_3AlSi_3O_{10}F_2$	Mica	Hazen and Finger (1978a)
Berndtite	SnS_2	Brucite	Hazen and Finger (1978b)

* Early single-crystal studies at the United States National Bureau of Standards (Block *et al.*, 1965; Weir *et al.*, 1965, 1969, 1971; Piermarini and Braun, 1972) are not included because full structure refinements were not performed. Structure refinements of *powders* at high pressure by time-of-flight neutron diffraction (Jorgenson, 1978; Srinivasa *et al.*, 1977, 1979; Cartz *et al.*, 1979) are also omitted.

Data on bond compressibilities from both fixed-structure compounds and variable structures are compiled in Table 7-2 (after Hazen and Finger, 1979a). Note that only about 100 different compounds are represented. Even so, these data have led to several generalizations regarding the variation of crystal structure with pressure.

B.   Empirical bond distance–pressure relationships*

Percy Bridgman, a great pioneer in high-pressure physics, was perhaps the first to attempt an empirical expression to allow the prediction of crystal bulk moduli. In his classical study of the compression of 30 metals, Bridgman (1923) found that compressibility was proportional to the 4/3rds power of molar volume. The importance of mineral bulk moduli in modelling the solid earth led Orson Anderson and his coworkers (Anderson and Nafe, 1965; Anderson and Anderson, 1970; Anderson, 1972) to adapt Bridgman's treatment to mineral-like compounds. For isostructural materials, it is found that compressibility is proportional to molar volume, or, as expressed in Anderson's papers:

$$\text{Bulk modulus} \times \text{molar volume} = \text{constant} \qquad (7\text{-}10)$$

* This section is derived from Hazen and Finger (1979).

Table 7-2. Polyhedral bulk moduli, bond distances, and bonding parameters for a variety compounds

Compound (polyhedron)	Structure Type	$\langle d \rangle$†, Å	K_p,‡ Mbar	z_c	z_a	S^2	$\dfrac{K_p \langle d \rangle^3}{S^2 z_c z_a}$	Reference
NiO	NaCl*	2.084	1.96(10)	2	2	0.50	8.9	1
MgO	NaCl*	2.106	1.61(5)	2	2	0.50	7.5	2
CoO	NaCl*	2.133	1.85(9)	2	2	0.50	9.0	1
FeO	NaCl*	2.139	1.53(8)	2	2	0.50	7.5	3
MnO	NaCl*	2.222	1.43(7)	2	2	0.50	7.8	1
CaO	NaCl*	2.406	1.10(5)	2	2	0.50	7.7	1
SrO	NaCl*	2.580	0.91(5)	2	2	0.50	7.8	4
BaO	NaCl*	2.770	0.69(4)	2	2	0.50	7.4	5
BeO	Zincite	1.66	2.5(5)	2	2	0.50	5.7	6
ZnO	Zincite	1.80	1.4(3)	2	2	0.50	4.2	6
UO_2	Fluorite*	2.37	2.3(1)	4	2	0.50	7.7	7
ThO_2	Fluorite*	2.42	1.93(10)	4	2	0.50	6.8	6
Al_2O_3	Corundum	1.91	2.4(2)	3	2	0.50	5.6	8
Fe_2O_3	Corundum	1.98	2.3(3)	3	2	0.50	6.0	8
Cr_2O_3	Corundum	1.99	2.3(3)	3	2	0.50	6.1	8
V_2O_3	Corundum	2.01	1.8(3)	3	2	0.50	4.7	8
SiO_2	Rutile§	1.778	3.2(1.5)	4	2	0.50	4.4	9
GeO_2	Rutile	1.884	2.7(1.0)	4	2	0.50	4.4	7
TiO_2	Rutile	1.961	2.2(1.0)	4	2	0.50	4.2	7
RuO_2	Rutile	1.968	2.7(1.0)	4	2	0.50	5.1	7
MnO_2	Rutile§	1.88	2.8(1.0)	4	2	0.50	4.7	9
SnO_2	Rutile	2.054	2.3(1.0)	4	2	0.50	4.9	7
SiO_2	Quartz¶	1.61	>5	4	2	0.50	>5.3	10
GeO_2	Quartz¶	1.73	>4	4	2	0.50	>4.9	10
Pyrope (Mg–O)	Garnet	2.27	1.3(1)	2	2	0.50	7.6	11
Pyrope (Al–O)	Garnet	1.89	2.2(5)	3	2	0.50	5.0	11
Pyrope (Si–O)	Garnet	1.63	3(1)	4	2	0.50	3.2	11
Grossular (Ca–O)	Garnet	2.40	1.15(13)	2	2	0.50	7.9	11
Grossular (Al–O)	Garnet	1.93	2.2(5)	3	2	0.50	5.3	11
Grossular (Si–O)	Garnet	1.64	3(1)	4	2	0.50	3.3	11
Forsterite (Mg–O)	Olivine	2.12	1.5(3)	2	2	0.50	7.0	12
γ-Ni_2SiO_4 (Ni–O)	Spinel	2.06	1.5(3)	2	2	0.50	6.6	13
γ-Ni_2SiO_4 (Si–O)	Spinel¶	1.66	>2.5	4	2	0.50	>2.9	13
Fassaite (Ca–O)	Clinopyroxene	2.49	0.85(20)	2	2	0.50	6.6	14
Phlogopite (K–O)	Mica	2.97	0.27(6)	1	2	0.50	7.1	15
Albite (Na–O)	Feldspar	2.75	0.32(6)	1	2	0.50	6.7	16
Zircon (Zr–O)	Zircon	2.20	2.8(3)	4	2	0.50	7.5	17
Zircon (Si–O)	Zircon¶	1.61	>2.5	4	2	0.50	>2.6	17
LiF	NaCl*	2.023	0.66(3)	1	1	0.75	7.0	18
NaF	NaCl*	2.310	0.45(2)	1	1	0.75	7.4	18
KF	NaCl*	2.674	0.293(15)	1	1	0.75	7.5	18
RbF	NaCl*	2.820	0.273(14)	1	1	0.75	8.2	6
LiCl	NaCl*	2.565	0.315(16)	1	1	0.75	7.0	6
NaCl	NaCl*	2.814	0.240(12)	1	1	0.75	7.1	6
KCl	NaCl*	3.146	0.180(9)	1	1	0.75	7.5	6
RbCl	NaCl*	3.291	0.160(8)	1	1	0.75	7.6	6
LiBr	NaCl*	2.750	0.257(13)	1	1	0.75	7.1	6
NaBr	NaCl*	2.989	0.200(10)	1	1	0.75	7.1	6
KBr	NaCl*	3.264	0.152(8)	1	1	0.75	7.0	6
RbBr	NaCl*	3.427	0.138(7)	1	1	0.75	7.4	6
LiI	NaCl*	3.000	0.188(9)	1	1	0.75	6.7	6
NaI	NaCl*	3.236	0.161(8)	1	1	0.75	7.3	6

Table 7-2 *continued*

Compound (polyhedron)	Structure Type	$\langle d \rangle$† Å	K_p,‡ Mbar	z_c	z_a	S^2	$\dfrac{K_p \langle d \rangle^3}{S^2 z_a z_a}$	References**
KI	NaCl*	3.533	0.124(6)	1	1	0.75	7.3	6
RbI	NaCl*	3.671	0.111(6)	1	1	0.75	7.5	6
CsCl	CsCl*	3.57	0.182(9)	1	1	0.75	11.1	18
CsBr	CsCl*	3.71	0.155(8)	1	1	0.75	10.6	6
CsI	CsCl*	3.95	0.129(6)	1	1	0.75	10.6	6
ThCl	CsCl*	3.32	0.236(12)	1	1	0.75	11.5	6
ThBr	CsCl*	3.43	0.225(11)	1	1	0.75	12.0	6
CuCl	Cubic ZnS*	2.34	0.40(2)	1	1	0.75	6.8	6
AgI	Cubic ZnS*	2.80	0.243(12)	1	1	0.75	7.1	3
CaF_2	Fluorite*	2.36	0.86(4)	2	1	0.75	7.6	6
BaF_2	Fluorite*	2.68	0.57(3)	2	1	0.75	7.3	6
PbF_2	Fluorite*	2.57	0.61(3)	2	1	0.75	6.9	6
SrF_2	Fluorite*	2.51	0.70(4)	2	1	0.75	7.4	6
MgF_2	Rutile§	1.99	1.0(3)	2	1	0.75	5.3	6
MnF_2	Rutile	2.12	0.9(2)	2	1	0.75	6.0	19
TaC	NaCl*	2.227	2.2(2)	4	4	0.2	7.6	6
TiC	NaCl*	2.159	1.9(2)	4	4	0.2	6.0	6
UC	NaCl*	2.480	1.6(1)	4	4	0.2	7.6	6
ZrC	NaCl*	2.341	1.9(2)	4	4	0.2	7.6	6
C	Diamond*	1.544	5.8(3)	4	4	0.2	6.7	6
C (planar C–C)	Graphite‖	1.42	5.9(3)	4	4	0.2	5.3	20
CaS	NaCl*	2.845	0.43(2)	2	2	0.4	6.2	3
SrS	NaCl*	3.010	0.40(2)	2	2	0.4	6.8	3
BaS	NaCl*	3.194	0.35(2)	2	2	0.4	7.2	3
PbS	NaCl*	2.968	0.48(3)	2	2	0.4	7.8	3
CdS	Zincite	2.43	0.61(6)	2	2	0.4	5.5	6
ZnS	Zincite	2.24	0.77(8)	2	2	0.4	5.5	6
ZnS	Cubic ZnS*	2.34	0.76(4)	2	2	0.4	6.1	6
SnS_2	CdI_2	2.56	1.2(2)	4	2	0.4	6.3	15
ZnSe	Cubic ZnS*	2.45	0.60(3)	2	2	0.4	5.5	6
CaSe	NaCl*	2.960	0.49(2)	2	2	0.4	7.9	3
SrSe	NaCl*	3.115	0.43(2)	2	2	0.4	8.1	3
BaSe	NaCl*	3.300	0.36(2)	2	2	0.4	8.1	3
PbSe	NaCl*	3.062	0.34(2)	2	2	0.4	6.1	3
CdSe	Zincite	2.35	0.54(5)	2	2	0.4	4.4	6
CaTe	NaCl*	3.178	0.42(2)	2	2	0.4	8.4	3
SrTe	NaCl*	3.235	0.334(16)	2	2	0.4	7.1	3
BaTe	NaCl*	3.493	0.305(15)	2	2	0.4	8.1	3
PbTe	NaCl*	3.227	0.41(2)	2	2	0.4	8.6	6
SnTe	NaCl*	3.157	0.42(2)	2	2	0.4	8.4	6
CdTe	Cubic ZnS*	2.81	0.42(2)	2	2	0.4	5.9	6
HgTe	Cubic ZnS*	2.78	0.44(2)	2	2	0.4	5.9	6
ZnTe	Cubic ZnS*	2.64	0.51(3)	2	2	0.4	5.9	6
GaSb	NaCl*	3.059	0.56(3)	3	3	0.25	7.1	6
InSb	NaCl*	3.239	0.47(2)	3	3	0.25	7.1	6
GaAs	NaCl*	2.827	0.75(4)	3	3	0.25	7.5	6
InAs	NaCl*	3.018	0.58(3)	3	3	0.25	7.1	6
GaP	NaCl*	2.726	0.89(4)	3	3	0.25	8.0	6
InP	NaCl*	2.934	0.73(4)	3	3	0.25	8.1	6
BN (linear B–N)	Boron nitrate‖	1.45	4.4(2)	3	3	0.2	7.5	20
KCN	NaCl*	3.263	0.143(7)	1	1	0.75	6.6	6

Notes overleaf

* Fully-constrained structures in which the polyhedral bulk modulus is identical to the crystal bulk modulus.
† Errors in bond distances at room pressures at $\leqslant 0.005$ Å.
‡ Errors in polyhedral bulk moduli are assumed to be $\pm 5\%$ for fully-constrained structures, unless worse precision is reported. Errors in polyhedral bulk moduli of unconstrained structures are generally greater than 15%.
§ High-pressure crystal structures of rutile-type SiO_2, MnO_2 and MgF_2 have not been done. Studies of other rutile-type compounds (references 7 and 19) indicate that polyhedral bulk moduli agree within $\pm 20\%$ of crystal bulk moduli.
‖ In graphite and boron nitride bond compression is constrained to be equal to unit-cell compression within the (001) plane.
¶ Tetrahedral bulk moduli of these compounds are greater than the indicated value. No upper limit is reported because net compression of the bond is comparable to the standard error of the bond distance.
** References to Table 7.2: 1, Clendenen and Drickamer (1966); 2, Schreiber and Anderson (1966); 3, Birch (1966); 4, Liu and Bassett (1973); 5, Liu and Bassett (1972); 6, Simmons and Wang (1971); 7, Hazen and Finger (1981); 8, Sato and Akimoto (1979) and Finger and Hazen (1980); 9, Bassett and Takahashi (1974); 10, Jorgensen (1978); 11, Hazen and Finger (1978c); 12, Hazen (1976a); 13, Finger *et al.* (1979); 14, Hazen and Finger (1977); 15, Hazen and Finger (1978a); 16, Hazen and Prewitt (1977b); 17, Hazen and Finger (1979c); 18, Yagi (1978); 19, Hazen *et al.* (1978); 20, Lynch and Drickamer (1966).

Typical bulk modulus–volume (KV) plots are shown in Figure 7-1. Although this relationship is empirical, theoretical arguments in support of constant KV may be derived from the simple two-term bonding potential expression of equation (7-9) (Anderson, 1972).

The same theoretical arguments used to explain the observed KV relationship in isostructural compounds may be used to predict a bulk modulus–volume relationship for cation coordination polyhedra. Hazen and Prewitt (1977a) found such an empirical trend in cation polyhedra from oxides and silicates:

$$K_p d^3 / z_c = \text{constant} \tag{7-11}$$

where z_c is the cation formal charge, d is the cation–anion mean bond distance, and K_p is the polyhedral bulk modulus. Polyhedral bulk moduli are easily calculated from the mean bond compressibility data in Table 7-2: $K = (3\beta_j)^{-1}$. Note that an 'effective' polyhedral bulk modulus may be calculated for planar and linear groups with this equation.

An important observation, based on data in Table 7-2, is that a given type of polyhedron has nearly constant bulk modulus within estimated experimental error, *independent* of structure. This constancy of polyhedral bulk modulus is especially remarkable because individual bonds within a polyhedron may show a wide range of compressibilities. Magnesium–oxygen octahedra in MgO, orthosilicates, layer silicates and chain silicates have the same bulk moduli (Figure 7-2). Errors associated with polyhedral bulk moduli are often in excess of 10 per cent, because bulk moduli are based on the *differences* between similar bond lengths. Even so, it appears that most polyhedra of a given type have bulk moduli that agree within ± 15 per cent of each other. If polyhedral bulk moduli are independent of structure, then the high-pressure behaviour of numerous materials, including glasses and other structures that are difficult to analyse, may be predicted.

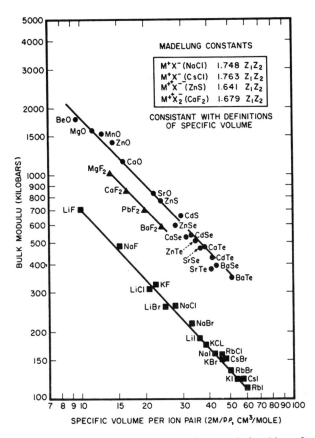

Figure 7-1. The bulk modulus–volume relationship of Anderson and coworkers (after Anderson, 1972). The product of bulk modulus and molar volume for isostructural compounds is approximately constant (the slopes of the solid lines are −1). (Reproduced by permission of McGraw-Hill Book Company)

The constancy of polyhedral bulk moduli may be questioned in the case of silicon tetrahedra. High-pressure single-crystal x-ray experiments have been limited to about 60 kbar, with a precision on Si–O distance no greater than 0.002 A. It is not possible, therefore, to determine accurately the large (>2.0 Mbar) bulk moduli of silicon tetrahedra. Although all authors agree that silicon tetrahedra are very rigid compared to monovalent or divalent cation polyhedra, it is not obvious whether all silicon tetrahedra have the same bulk modulus. These tetrahedra exhibit perhaps the greatest range of linkage topologies of any common polyhedron, with framework, chain, sheet, ring and isolated tetrahedral groups. Tetrahedra in several of these topologies may undergo bending of T–O–T bonds as well as bond compression. It has been suggested (M. Vaughan, personal communication) that in silicates which experience extensive T–O–T bending at high

156

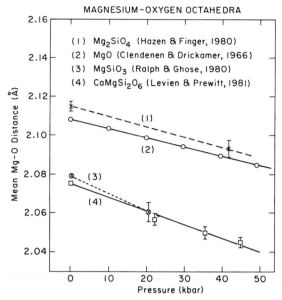

Figure 7-2. Magnesium–oxygen bond distances versus pressure for octahedra in several compounds

pressure the tetrahedral bulk moduli will be greater than in structures with little or no bending, because the energy applied to bond bending cannot be applied to bond compression as well. If this is so then silicon polyhedral bulk moduli will be different in different structures, according to the degree of flexibility of the tetrahedral network. At present, however, the bulk moduli of silicon tetrahedra cannot be measured with sufficient accuracy to confirm this plausible hypothesis.

Consider the 38 polyhedral bulk moduli that may be derived from data in Table 7-2 on oxides and silicates. A relationship, analogous to that proposed by Hazen and Prewitt (1977a; see equation 7-11) is obtained:

$$\frac{K_p d^3}{z_c} = 7.5 \pm 0.2 \text{ Mbar-A}^3 \qquad (7\text{-}12)$$

where K_p is the polyhedral bulk modulus in Mbar, d is the mean cation–anion distance in A, and z_c is the integral formal charge of the cation. Polyhedral data used to derive equation (7-12) are illustrated in Figure 7-3. The coefficient is calculated from weighted linear regression of data in Table 7-2, and was constrained to pass through the origin. Calculations were performed using both K_p^{-1} and d^3/z_c as independent variables; 7.5 $\pm$ 0.2 is the average coefficient. Weighted regression of K_p versus z_c/d^3 yields a slightly smaller coefficient of 7.3 $\pm$ 0.2. Bulk moduli from fully constrained structures, notably NaCl-type oxides, are known more precisely than other polyhedral bulk moduli; the coefficient is, therefore, heavily weighted by these oxides. Almost all data, however, lie within two estimated standard deviations of the predicted bulk moduli.

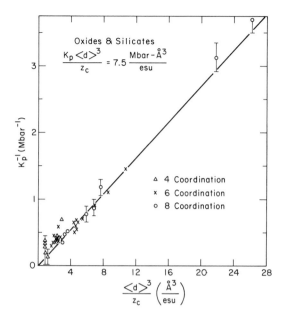

Figure 7-3. The bulk modulus–volume relationship for polyhedra in oxides and silicates. Triangles represent tetrahedra; crosses, octahedra; and circles, eight- or greater-coordination sites. Error bars represent one estimated standard deviation in polyhedral bulk moduli for selected polyhedra in silicates. The line is a weighted linear-regression fit of all data, constrained to pass through the origin. This line is largely determined by polyhedra in fixed-structure oxides, for which polyhedral bulk modulus errors are small. Almost all points, however, lie within two estimated standard deviations of predicted bulk moduli. (From Hazen and Finger, 1979a, reproduced by permission of the American Geophysical Union)

Equation (7-12) is consistent with observed high-pressure crystal structure refinements (see section II-A-2). Bond compressibility is proportional to the cube of the mean bond distance in a polyhedron, and inversely proportional to cation formal charge. Thus large alkali cation polyhedra are much more compressible than small Al^{3+} or Si^{4+} tetrahedra. Divalent cation octahedra have an intermediate compressibility.

A similar bulk modulus–volume relationship for halides may be derived from data in Table 7-2:

$$\frac{K_p d^3}{z_c} = 5.6 \pm 0.1 \text{ Mbar-A}^3 \qquad (7-13)$$

Equations (7-12) for oxygen-based compounds and (7-13) for halides may be combined with other data on sulphides, selenides, tellurides, phospides, arsenides, antimonides and carbides into a more general bulk modulus–volume relationship:

$$\frac{K_p d^3}{S^2 z_c z_a} = 7.5 \pm 0.2 \text{ Mbar-A}^3 \qquad (7\text{-}14)$$

where z_a is the formal anion charge and S^2 is an empirical term for the relative 'ionicity' of the bond, defined as 0.50 for R^{2+}–O bonds in NaCl-type oxides. Equation (7-14) is similar in form to equation (5) of Anderson (1972, p. 278), which is valid for the *bulk* properties of several simple compounds. When applied to the bulk properties of minerals, however, the constant may vary, depending on structure type. Furthermore, the appropriate value of $z_c z_a$ in many complex compounds is not obvious. In $NaNO_2$, for example, bulk modulus–volume data fit the alkali halide trend with $z_c z_a = 1$; this value does not apply to $NaNO_3$, however (Hazen and Finger, 1979b). These difficulties are not present when equation (7-14) is applied to cation coordination groups. A single constant appears to fit many different structure types, and the values of z_c and z_a are unambiguous.

Values of S^2 may be calculated from equations such as (7-12) and (7-13) for each type of anion. (Until a better measure of 'ionicity' is available it will be assumed that S^2 is constant for a given anion.) If S^2 is defined to be 0.50 for all oxides and silicates, then combining equations (7-13) and (7-14) yields S^2 for halides ≈ 0.75. Empirical values of S^2 determined in this way for other anions are 0.40 for sulphides, selenides and tellurides; 0.25 for phosphides, arsenides and antimonides; and 0.20 for carbides. Although the physical significance of S^2 is not obvious, the *same* values of S^2 are applicable to the empirical thermal expansion equation (6-18).

Of the oxide and silicate polyhedra used to construct Figure 7-3, those that deviate most from the empirical line are tetrahedra, such as Si in silicates and Zn in ZnO, and octahedrally coordinated vanadium in V_2O_3, an unusual oxide with metallic lustre and conductivity. All of these polyhedra, which are more compressible than predicted by equation (7-14), also have bonding that is more covalent than in the other plotted oxide and silicate polyhedra (i.e. S^2 may be less than 0.50.) Thus, deviations from the line in Figure 7-3 may provide an approximate measure of ionicity.

Figure 7-4 illustrates equation (7-14) for polyhedra in more than 100 substances in 19 different structure types from data in Table 7-2. It is significant that a simple empirical relationship successfully models bond compression in materials with a wide range of bond character and topology. Of the structures examined, only CsCl is anomalous; all four points representing that structure fall significantly below the empirical line. The CsCl structure, with eight anions at the corners of a unit cube, and a cation at the cube's centre, is unique in the high degree of face-sharing between adjacent polyhedra and the consequent short cation–cation separations. In CsCl-type compounds the cation–cation distance is only 15 per cent longer than cation–anion bonds, in contrast to the 50–75 per cent separation in most other structures. It is probable, therefore, that equation (7-14), which incorporates only the bonding characteristics of the primary

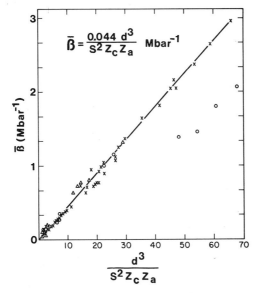

Figure 7-4. The bulk modulus–volume relationship for polyhedra in a variety of materials. Triangles, crosses and circles indicate coordination as in Figure 7-3. (Reproduced from Hazen and Finger, 1979a, by permission of the American Geophysical Union)

coordination sphere, is not valid for structures in which extensive polyhedral face-sharing results in strong second-nearest neighbour interactions.

C. Relationships between polyhedral and crystal bulk modulus

(1) Calculation of crystal bulk moduli

Crystal bulk modulus is a function of polyhedral linkages as well as polyhedral bulk moduli, as described in Chapter Six, section III-C. The pressure variation of a crystal structure is not generally the simple sum of polyhedral variations. Bond bending, which is not considered in the treatment of polyhedral bulk moduli, must also be taken into account. In compounds with significant bending of metal–oxygen–metal bonds, such as the quartz form of SiO_2, bulk compression may be many times greater than polyhedral compression. On the other hand, many compounds with simple or fixed structures, in which little metal–oxygen–metal bending can occur, have bulk moduli similar to those of component polyhedra. In order to predict the bulk modulus of a compound it is necessary to know the nature of polyhedral linkages, as well as the moduli of component polyhedra. Furthermore, data on bending force constants of bonds are also required. Bending force parameters are not now well known, although recent measurements of force constants may lead to prediction of polyhedral tilting magnitudes (G. V. Gibbs, personal communication.)

(2)   Pressure derivatives of bulk modulus

The bulk modulus of a crystal varies with pressure and, in general, as a crystal is compressed, K increases. Plots of pressure versus molar volume thus normally display a positive curvature. Precise pressure–volume data for alkali halides and other simple solids have been used to derive and test several P–V equations of state (EOS). The simplest EOS is a polynomial equation of the form:

$$V = V_0 + aP + bP^2 \qquad (7\text{-}15)$$

This EOS is not adequate, however, because a quadratic equation is not a physically reasonable model for high-pressure behaviour. The derivative of volume with respect to pressure must be negative, but a parabolic form has positive slope at high pressure.

More realistic are the Birch EOS:

$$P = \frac{K_0}{K_0'} \left[\left(\frac{V_0}{V} \right)^{K_0'} - 1 \right] \qquad (7\text{-}16)$$

and the Birch–Murnaghan EOS:

$$P = \frac{3}{2} K_0 \left[\left(\frac{V_0}{V} \right)^{7/3} - \left(\frac{V_0}{V} \right)^{5/3} \right] \left\{ 1 - \frac{3}{4}(4 - K_0') \left[\left(\frac{V_0}{V} \right)^{2/3} - 1 \right] \right\} \quad (7\text{-}17)$$

These two *empirical* EOS, although considerably different in form, yield equivalent results for bulk modulus and its pressure derivative at compression of $\Delta V/V < 0.15$. For greater compression, or very precise P–V data, a higher-order EOS of a different form may be required (Chhabildas and Ruoff, 1976).

It is logical to assume that the pressure–distance relationship for bonds in a crystal is also a complicated function, with P–d EOS analogues to equations (7-16) and (7-17). In fact, for fixed structures such as NaCl and CaF$_2$, bond compression must have precisely the same form as the P–V EOS. High-pressure crystal structure refinements, at present, are not of sufficient precision to detect curvature in most cation–anion bond distances as a function of pressure. The use of a constant K_p for polyhedral bulk moduli below 100 kbar is thus employed in subsequent sections of this book.

III.   OTHER STRUCTURAL VARIATIONS WITH PRESSURE

A.   Polyhedral distortions

Hazen and Finger (1978c) suggested that at higher pressure cation polyhedra may become more regular and polyhedral distortion indices closer to their ideal values. This hypothesis was based on the observed behaviour of cation polyhedra in olivines and garnets, in which longer metal–oxygen bonds compress more than short metal–oxygen bonds. Subsequent studies of zircon (Hazen and Finger, 1979c), corundum-type compounds (Finger and Hazen, 1980), and rutile-type oxides (Hazen and Finger, 1981), however, reveal the opposite trend, with polyhedra becoming more distorted at high pressure. It appears, therefore, that no general trends exist in the variation of polyhedral distortion with pressure.

B. Variation of temperature factors with pressure

Finger and King (1978) have demonstrated that pressure has a small, but possibly measurable, effect on the isotropic temperature factor. The average energy, E, associated with a vibrating bond of mean ionic separation, d, and mean-square displacement, $\langle r^2 \rangle$ ($r \ll d$), is:

$$E \approx \frac{z_c z_a e^2}{2d^3}(ad - 2)\langle r^2 \rangle \tag{7-18}$$

where z_c and z_a are cation and anion charges, and a is a repulsion parameter (Karplus and Porter, 1970). The isotropic temperature factor, B, is proportional to the mean-square displacement:

$$B = 8\pi^2 \langle r^2 \rangle. \tag{7-19}$$

Therefore, combining equations (7-18) and (7-19),

$$E = \frac{z_c z_a e^2 B}{16\pi^2 d^3}(ad - 2) \tag{7-20}$$

If it is assumed that the average energy, E, and the repulsion parameter, a, are independent of pressure, then the temperature factor at pressure, B_p, is related to the room-pressure temperature factor, B_0, as follows:

$$B_p = B_0 \frac{(ad_0 - 2)d_p^3}{(ad_p - 2)d_0^3} \tag{7-21}$$

In the case of NaCl at 32 kbar, Finger and King (1978) predicted a 5.7 per cent reduction in the temperature factors of Na and Cl at high pressure. The observed reductions of approximately 10 per cent (± 5 per cent) provided evidence for the proposed effect of pressure on amplitude of atomic vibrations.

IV. CONCLUSIONS

Two important conclusions have been reached regarding the variation of crystal structure with pressure. First, as with temperature, a given change of pressure appears to have an approximately constant effect on the size of a cation coordination polyhedron, independent of structural linkages. Second, the magnitude of this effect is given by the bulk modulus–volume relationship for cation polyhedra:

$$K_p = 7.5 \ (\pm 0.2) \ S^2 z_c z_a / d^3$$

This relationship may require modification, however, to take the effects of bond bending into account.

162

The bulk modulus–volume relationship for cation coordination polyhedra is a useful *empirical* equation for predicting the geometry of compression in many solids. The relationship may also be used to predict approximate macroscopic bulk moduli in some fully linked structures, although uncertainties in the magnitudes of polyhedral tilting and in the ionicity coefficients of different bonds limit this application. Equation (7-14), therefore, is not intended as a substitute for the successful bulk modulus–volume relationships of Anderson (1972) and others.

Equation (7-14) places significant constraints on theoretical models of compression and bonding (see equation 7-9), although equation (7-14) is not an explanation of cation–anion compression. The empirical bulk modulus–volume relationship should be considered in the formulation of models of interatomic forces, as well as in the description and prediction of crystalline compression.

REFERENCES

Anderson, D. L., and O. L. Anderson (1970) The bulk modulus–volume relationship for oxides. *J. Geophys. Res.* **75**, 3494–3500.

Anderson, O. L. (1972) Patterns in elastic constants of minerals important to geophysics. In *Nature of the Solid Earth*, E. C. Robinson (ed.), McGraw-Hill, New York, 575–613.

Anderson, O. L., and J. E. Nafe (1965) The bulk modulus–volume relationship for oxide compounds and related geophysical problems. *J. Geophys. Res.*, **70**, 3951–3963.

Bassett, W. A., and T. Takahashi (1974) X-ray diffraction studies up to 300 kbar. *Adv. High Pressure Res.* **4**, 165–247.

Birch, F. (1966) Compressibility: Elastic constants. In *Handbook of Physical Constants, Geol. Soc. Am. Memoir* **97**, 97–173.

Block, S., C. E. Weir and G. J. Piermarini (1965) High pressure single crystal studies of Ice VI. *Science* **148**, 947–948.

Bridgman, P. W. (1923) The compressibility of thirty metals as a function of temperature and pressure. *Proc. Am. Acad. Arts Sci.* **58**, 165–242.

Bukowinski, M. S. T. (1980) Effect of pressure on bonding in MgO. *J. Geophys. Res.* **85**, 285–292.

Cartz, L., S. R. Srinivasa, R. J. Riedner, J. D. Jorgensen and T. G. Worlton (1979) Effect of pressure on bonding in black phosphorus. *J. Chem. Phys.* **71**, 1718–1721.

Chhabildas, L. C., and A. L. Ruoff (1976) Isothermal equation of state for sodium chloride by the length-change-measurement technique. *J. Appl. Phys.* **47**, 4182–4187.

Clendennen, R. L., and H. G. Drickamer (1966) Lattice parameters of nine oxides and sulfides as a function of pressure. *J. Chem. Phys.* **44**, 4223–4228.

D'Amour, H., W. Denner and H. Schulz (1979) Structure determination of α-quartz up to 68×10^8 Pa. *Acta Crystallogr.* **B35**, 550–555.

D'Amour, H., D. Schiferl, W. Denner, H. Schulz and W. B. Holzapfel (1978) High-pressure single-crystal structure determinations for ruby up to 90 kbar using an automatic diffractometer. *J. Appl. Phys.* **49**, 4411–4416.

Das, C. D., H. V. Keer and R. V. G. Rao (1963) Lattice energy and other properties of some ionic crystals. *Z. Physik. Chem.* **224**, 377–383.

Denner, W., H. Schulz and H. D'Amour (1979) The influence of high hydrostatic pressure on the crystal structure of cesium gold chloride ($Cs_2Au^IAu^{III}Cl_6$) in the pressure range up to 52×10^8 Pa. *Acta Cryst.* **A35**, 360–365.

Finger, L. W., and R. M. Hazen (1978) Crystal structure and compression of ruby to 46 kbar. *J. Appl. Phys.* **49**, 5823–5826.

Finger, L. W., and R. M. Hazen (1980) Crystal structure and isothermal compression of Fe_2O_3, Cr_2O_3, and V_2O_3 to 50 kbar. *J. Appl. Phys.* **51**, 5362–5367.

Finger, L. W., R. M. Hazen and T. Yagi (1977) High pressure crystal structures of the spinel polymorphs of Fe_2SiO_4 and Ni_2SiO_4. *Carnegie Inst. Washington Year Book* **76**, 504–505.

Finger, L. W., R. M. Hazen and T. Yagi (1979) Crystal structures and electron densities of nickel and iron silicate spinels at elevated temperature or pressure. *Am. Mineral.* **64**, 1002–1009.

Finger, L. W., and H. King (1978) A revised method of operation of the single-crystal diamond cell and refinement of the structure of NaCl at 32 kbar. *Am. Mineral.* **63**, 337–342.

Hazen, R. M. (1975) *Effects of Temperature and Pressure on the Crystal Physics of Olivine*, Ph.D. Thesis, Department of Geological Sciences, Harvard University.

Hazen, R. M. (1976a) Effects of temperature and pressure on the crystal structure of forsterite. *Am. Mineral.* **61**, 1280–1293.

Hazen, R. M. (1976b) Effects of temperature and pressure on the cell dimension and x-ray temperature factor of periclase. *Am. Mineral.* **61**, 266–271.

Hazen, R. M. (1977) Effects of temperature and pressure on the crystal structure of ferromagnesian olivine. *Am. Mineral.* **62**, 286–295.

Hazen, R. M., and C. W. Burnham (1974) The crystal structures of gillespite I and II: a structure determination at high pressure. *Am. Mineral.* **59**, 1166–1176.

Hazen, R. M., and C. W. Burnham (1975) The crystal structure of gillespite II at 26 kilobars: correction and addendum. *Am. Mineral.* **60**, 937–938.

Hazen, R. M., and L. W. Finger (1977) Compressibility and crystal structure of Angra dos Reis Fassaite to 52 kbar. *Carnegie Inst. Washington Year Book* **76**, 512–515.

Hazen, R. M., and L. W. Finger (1978a) The crystal structures and compressibilities of layer minerals at high pressure. II. Phlogopite and chlorite. *Am. Mineral.* **63**, 293–296.

Hazen, R. M., and L. W. Finger (1978b) The crystal structures and compressibilities of layer minerals at high pressure. I. SnS_2, berndtite. *Am. Mineral.* **63**, 289–292.

Hazen, R. M., and L. W. Finger (1978c) Crystal structures and compressibilities of pyrope and grossular to 60 kbar. *Am. Mineral.* **63**, 297–303.

Hazen, R. M., and L. W. Finger (1979a) Bulk modulus-volume relationship for cation-anion polyhedra. *J. Geophys. Res.* **84**, 6723–6728.

Hazen, R. M., and L. W. Finger (1979b) Linear compressibilities of $NaNO_2$ and $NaNO_3$. *J. Appl. Phys.* **50**, 6826–6828.

Hazen, R. M., and L. W. Finger (1979c) Crystal structure and compressibility of zircon at high pressure. *Am. Mineral.* **64**, 196–201.

Hazen, R. M., and L. W. Finger (1980) Crystal structure of forsterite at 40 kbar. *Carnegie Inst. Washington Year Book* **79**, 364–367.

Hazen, R. M., and L. W. Finger (1981) Bulk moduli and high-pressure crystal structures of rutile-type compounds. *J. Phys. Chem. Solids* **42**, 143–151.

Hazen, R. M., L. W. Finger and T. Yagi (1978) Crystal structure and compressibility of MnF_2 to 15 kbar. *Carnegie Inst. Washington Year Book* **77**, 841–842.

Hazen, R. M., and C. T. Prewitt (1977a) Effects of temperature and pressure on interatomic distances in oxygen-based minerals. *Am. Mineral.* **62**, 309–315.

Hazen, R. M., and C. T. Prewitt (1977b) Linear compressibilities of low albite: high-pressure structural implications. *Amer. Mineral.* **62**, 554–558.

Jorgensen, J. D. (1978) Compression mechanisms in α-quartz structures—SiO_2 and GeO_2. *J. Appl. Phys.* **49**, 5473–5478.

Karplus, M., and R. N. Porter (1970) *Atoms and Molecules: An Introduction for Students of Physical Chemistry*, W. A. Benjamin, Inc., Menlo Park, California.

Keller, R., W. B. Holzapfel and H. Schulz (1977) Effect of pressure on atom positions in Se and Te. *Phys. Rev. B* **16**, 4404–4412.

King, H. E. (1978) Compression in sulfide crystal structures: a comparison with oxides and silicates. *Geol. Soc. Am. Abstracts with Programs* **10**, 434.

Levien, L., and C. T. Prewitt (1981a) High-pressure structural study of diopside. *Am. Mineral.* **66**, 315–323.

Levien, L., and C. T. Prewitt (1981b) High-pressure crystal structure and compressibility of coesite. *Am. Mineral.* **66**, 324–333.

Levien, L., C. T. Prewitt and D. J. Weidner (1980) Structure and elastic properties of quartz at pressure. *Am. Mineral.* **65**, 920–930.

Liu, L. G., and W. A. Bassett (1972) Effect of pressure on the crystal structure and the lattice parameters of BaO. *J. Geophys. Res.* **77**, 4934–4937.

Liu, L. G., and W. A. Bassett (1973) Changes of the crystal structure and the lattice parameter of SrO at high pressure. *J. Geophys. Res.* **78**, 8470–8473.

Lynch, R. W., and H. G. Drickamer (1966) Effect of high pressure on the lattice parameters of diamond, graphite, and hexagonal boron nitride. *J. Chem. Phys.* **44**, 181–184.

Merrill, L., and W. A. Bassett (1975) The crystal structure of $CaCO_3$ II, a high-pressure metastable phase of calcium carbonate. *Acta Crystallogr.* **B31**, 343–349.

Piermarini, G. J., and A. B. Braun (1972) Crystal and molecular structure of CCl_4 III: A high pressure polymorph at 10 kbar. *J. Chem. Phys.* **58**, 1974–1982.

Ralph, R., and S. Ghose (1980) Enstatite, $Mg_2Si_2O_6$: compressibility and crystal structure at 21 kbar. *Amer. Geophys. Union Trans (EOS)* **61**, V174.

Sato, Y., and S. Akimoto (1979) Hydrostatic compression of four corundum-type compounds: α-Al_2O_3, V_2O_3, Cr_2O_3, and α-Fe_2O_3. *J. Appl. Phys.* **50**, 5285–5291.

Schiferl, D. (1977) 50-kilobar gasketed diamond anvil cell for single-crystal x-ray diffractometer use with the crystal structure of Sb up to 26 kilobars as a test problem. *Rev. Scien. Instrum.* **48**, 24–30.

Schreiber, E., and O. L. Anderson (1966) Pressure derivatives of the sound velocity of polycrystalline alumina. *J. Am. Ceram. Soc.* **49**, 184–190.

Simmons, G., and H. Wang (1971) *Single Crystal Elastic Constants*, MIT Press, Cambridge, Massachusetts.

Srinivasa, S. R., L. Cartz, J. D. Jorgensen, T. G. Worlton, R. A. Beyerlein and M. Billy (1977) High-pressure neutron diffraction study of Si_2N_2O. *J. Appl. Cryst.* **10**, 167–171.

Srinivasa, S. R., L. Cartz, J. D. Jorgensen and L. C. Labbe (1979) Pressure induced tetrahedral tilting and deformation in Ge_2N_2O. *J. Appl. Cryst.* **12**, 511–516.

Tossell, J. A. (1980) Theoretical study of structures, stabilities, and phase transitions in some metal dihalide and dioxide polymorphs. *J. Geophys. Res.* **85**, 6456–6460.

Weir, C. E., S. Block and G. J. Piermarini (1965) Single crystal x-ray diffraction at high pressure. *J. Research Natl Bureau Standards (US)* **69C**, 275–281.

Weir, C. E., G. J. Piermarini and S. Block (1969) Crystallography of some high-pressure forms of C_6H_6, CS_2, Br_2, CCl_4, and KNO_3. *J. Chem. Phys.* **50**, 2089–2093.

Weir, C. E., G. J. Piermarini and S. Block (1971) On the crystal structure of Cs-II and Ga-II. *J. Chem. Phys.* **54**, 2768–2770.

Yagi, T. (1978) Experimental determination of thermal expansivity of several alkali halides at high pressures. *J. Phys. Chem. Solids* **39**, 563–571.

Chapter 8

Structural Variations with Composition

CONTENTS

I   Introduction 165
II   Effects of Composition on Structural Dimensions 166
     A   Ionic Radii 166
     B   Effects of Composition on Molar Volume 168
         (1) Vegard's law 168
         (2) Excess volume of mixing 171
     C Effects of Composition on Bond Distance and Polyhedral Volume . . . 173
III Conclusions 175

I. INTRODUCTION

Composition is an internal variable—a characteristic intrinsic to each crystal studied, and is distinct from the variables of temperature and pressure, which may be imposed almost instantaneously on a crystal by external means. Studies of the effects of compositional variation on a crystal structure, therefore, require a different crystal for each composition of interest. Nevertheless, the continuous variation of crystal structure with composition may be quantified, just as with changes of temperature or pressure.

A useful definition is the coefficient of compositional expansion, γ:

$$\text{linear:} \; \gamma_l = \frac{1}{d}\left(\frac{\partial d}{\partial X}\right)_{T,P} \tag{8-1}$$

$$\text{volume:} \; \gamma_v = \frac{1}{V}\left(\frac{\partial V}{\partial X}\right)_{T,P} \tag{8-2}$$

Equations (8-1) and (8-2) have precisely the same form as equations (6-11) and

(6-12) for α and equations (7-1) and (7-2) for β. A useful convention in applying equations (8-1) and (8-2) is that if two cations are in solid solution, X refers to the mol fraction of the larger cation. Thus, the coefficient of volume expansion is always positive. The units of γ are fractional change in distance or volume per mol.

II. EFFECTS OF COMPOSITION ON STRUCTURAL DIMENSIONS

A. Ionic radii

The principal continuous variations of structure that result from changes in composition are due to differences in electronic structure between the original and substituting atoms. In addition to obvious effects on site electron density, these differences, as manifest in differing ionic radii, affect volume and distance parameters of the crystal. Many structural variations resulting from compositional substitution, therefore, may be understood in terms of ionic radii.

Several authors, following the pioneering work of Bragg, Lande, Goldschmidt and others (see Pauling, 1960), have proposed internally consistent sets of radii as a function of valence and coordination number. One of the most widely quoted radii tables was developed by Shannon and Prewitt (1970) and revised by Shannon (1976). All radii tables require the assumption of one standard radius, because the values observed in an experiment are interatomic distances. In the Shannon and Prewitt compilation the radius of oxygen was set at 1.40 A, in accord with the value chosen by Pauling. Selected values from the Shannon (1976) radii table are reproduced in Table 8-1.

A consequence of the 1.40 A value for oxygen radius is that anions are modelled as larger than cations in most ceramics and mineral-like compounds. Not all authors agree with the concept of 'small' cations, however. Slater (1963)

Table 8-1. Selected ionic radii (from Shannon, 1976)

Ion	Coordination	Radius	Ion	Coordination	Radius
Al^{3+}	IV	0.39	Ba^{2+}	X	1.52
	V	0.48		XI	1.57
	VI	0.535		XII	1.61
As^{3+}	VI	0.58	Be^{2+}	IV	0.27
As^{5+}	IV	0.335		VI	0.45
	VI	0.46	Br^{-1}	VI	1.96
B^{3+}	III	0.01	C^{4+}	III	−.08
	IV	0.11		IV	0.15
	VI	0.27	Ca^{2+}	VI	1.00
Ba^{2+}	VI	1.35		VII	1.06
	VII	1.38		VIII	1.12
	VIII	1.42		IX	1.18
	IX	1.47		X	1.23

Table 8-1 *continued*

Ion	Coordination	Radius	Ion	Coordination	Radius
Ca^{2+}	XII	1.34			
Cd^{2+}	VI	0.95	Na^+	IX	1.24
Ce^{4+}	VIII	0.97		XII	1.39
Cl^{-1}	VI	1.81	Ni^{2+}	IV	0.55
$Co^{2+}(HS)*$	VI	0.745		VI	0.690
$Cr^{2+}(HS)$	VI	0.80	O^{-2}	II	1.35
Cr^{3+}	VI	0.615		III	1.36
Cs^+	VI	1.67		IV	1.38
	VIII	1.74		VI	1.40
	XI	1.88		VIII	1.42
Eu^{2+}	VI	1.17	OH^-	II	1.32
	VIII	1.25		III	1.34
F^-	IV	1.31		IV	1.35
	VI	1.33		VI	1.37
$Fe^{2+}(HS)$	IV	0.63	P^{5+}	IV	0.17
	IV square	0.64		VI	0.38
(HS)	VI	0.780	Pb^{2+}	VI	1.19
	VIII	0.92		VIII	1.29
Fe^{3+}	IV	0.49	Rb^+	VI	1.52
(HS)	VI	0.645		VIII	1.61
Ga^{3+}	IV	0.47		IX	1.63
Ge^{4+}	IV	0.39		X	1.66
	VI	0.53		XII	1.72
H^+	I	−0.38	Ru^{4+}	VI	0.62
	II	−0.18	S^{2-}	VI	1.84
Hf^{4+}	VI	0.71	Se^{2-}	VI	1.98
	VIII	0.83	Si^{4+}	IV	0.26
Hg^{2+}	VI	1.02		VI	0.40
I^{-1}	VI	2.20	Sm^{3+}	VI	0.96
K^+	VI	1.38	Sn^{4+}	VI	0.69
	VIII	1.51	Sr^{2+}	VI	1.18
	IX	1.55		VIII	1.26
	X	1.59		XII	1.44
	XII	1.64	Te^{2-}	VI	2.21
Li^+	IV	0.59	Ti^{3+}	VI	0.670
	VI	0.76	Ti^{4+}	IV	0.42
	VIII	0.92		VI	0.605
Mg^{2+}	IV	0.57	U^{4+}	VI	0.89
	VI	0.720		VIII	1.00
	VIII	0.89	V^{2+}	VI	0.79
Mn^{2+} (HS)	VI	0.830	V^{3+}	VI	0.64
	VIII	0.96	V^{5+}	IV	0.355
Mn^{3+} (HS)	VI	0.645		VI	0.54
N^{3-}	IV	1.46	W^{4+}	VI	0.66
N^{3+}	VI	0.16	Zn^{2+}	IV	0.60
N^{5+}	III	−.10		VI	0.74
Na^+	VI	1.02	Zr^{4+}	VI	0.72
	VII	1.12		VIII	0.84
	VIII	1.18			

* (HS) denotes high-spin electronic configuration.

and Prewitt (1977) have suggested that smaller anions may be more realistic. O'Keeffe and Hyde (1981) have proposed, alternatively, that two sets of radii should be considered in the description of structure. Bonded radii are similar to the Pauling, and Shannon and Prewitt sizes, with anions larger than cations. In addition, O'Keeffe and Hyde propose the use of 'nonbonded' radii for next-nearest neighbour atoms. Nonbonded radii for cations are significantly greater than for anions. These second-nearest-neighbour radii have been used successfully to explain the near constant distance of some cation–cation pairs, such as Si–Si, in a wide variety of structures.

Regardless of the bonding-radii model employed, all sets of ionic radii yield the same *relative* sizes of cations. Cations of similar size and valence commonly substitute for each other in ionic compounds. In a few instances, such as solid solutions between zirconium and hafnium, two cations are so similar that the compositional changes have little measurable effect on structure. Generally, however, substituting cations differ by a few per cent in size. Systematic structure variations which result from such ionic substitution are examined below.

B. Effects of composition on molar volume

(1) Vegard's law

Vegard and Dale (1928) proposed the simple relationship that in isomorphous compounds, which differ only in ionic substitution on a specific site, molar volume is proportional to the cube of the radius of the substituting ion:

$$V \propto r^3 \tag{8-3}$$

The fundamental incorrectness of 'Vegard's Law' can be shown by considering the simple oxides, $R^{2+}O$, with the rock-salt structure, which include $R^{2+} = $ Ni, Mg, Co, Fe, Mn, Cd, Ca, Eu, Sr and Ba. In the rock-salt structure the unit-cell edge is twice the R–O distance; thus, the volume is given by

$$V = 8d_{R-O}^3 = 8\,(r_R + r_0)^3 \tag{8-4}$$

Obviously, a plot of volume versus the cube of the cation radius is not a straight line. However, a plot of the unit-cell edge (or $V^{1/3}$) versus cation radius is linear. Figure 8-1 is a classic Vegard's Law plot for the rock-salt structure oxides. The departure from linearity is not great in the range illustrated. Such plots, therefore, are useful in predicting molar volume of isomorphous compounds.

A more complex structure with which Vegard's Law may be examined is that of the trioctahedral micas, a common group of layer silicate minerals with the general formula $A^+ R_3^{2+} Y^{3+} Si_3 O_{10}(OH)_2$. The structure (Figure 8-2) consists of an octahedral R^{2+} layer sandwiched between two tetrahedral $Y Si_3$ layers. The three-layer sandwiches are linked by layers of A^+ alkali cations in the interlayer positions. Numerous chemical substitutions are known, including K, Na, Cs, Rb, and NH_4 for A^+; Ni, Mg, Fe, Co, Mn, and Cu for R^{2+}; and Al, Ga, Fe, and B for Y^{3+}. Both R^{2+} and Y^{3+} substitutions follow Vegard's Law (Figures 8-3 and 8-4). The

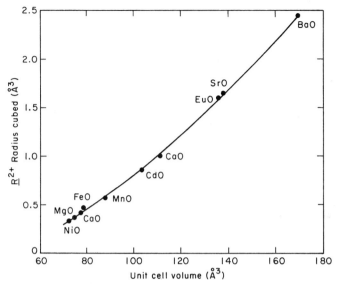

Figure 8-1. 'Vegard's Law' plot of unit-cell volume versus ionic radius cubed of R^{2+} for rock-salt type oxides ($R^{2+}O$). The 'Law' predicts a linear variation, though such a relationship obtains only for cations of similar radius

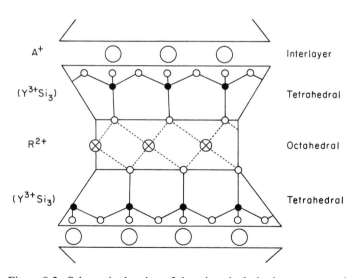

Figure 8-2. Schematic drawing of the trioctahedral mica structure. A layer of R^{2+} octahedra is sandwiched between two tetrahedral layers ($Y^{3+}Si_3$). These three-layer planar units are bound by alkali A^+ cations in the interlayer sites

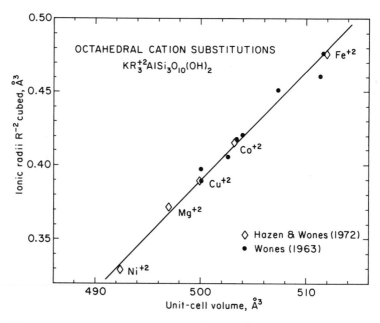

Figure 8-3. Unit-cell volume versus octahedral R^{2+} cation radius cubed for micas of the form $KR_3^{2+}AlSi_3O_{10}(OH)_2$. Diamonds from Hazen and Wones (1972), solid circles from Wones (1963). (Reproduced by permission of the Mineralogical Society of America)

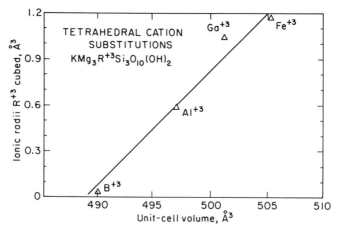

Figure 8-4. Unit-cell volume versus tetrahedral Y^{3+} cation radius cubed for micas of the form $KMg_3Y^{3+}Si_3O_{10}(OH)_2$. (From Hazen and Wones, 1972, reproduced by permission of the Mineralogical Society of America)

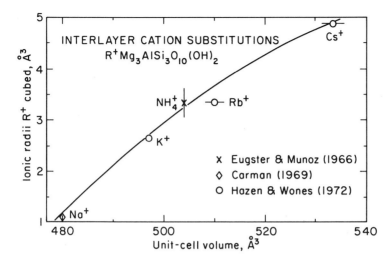

Figure 8-5. Unit-cell volume versus interlayer A^+ cation radius for micas of the form $A^+Mg_3AlSi_3O_{10}(OH)_2$. (From Hazen and Wones, 1972, reproduced by permission of the Mineralogical Society of America.)

alkali substitution, however, appears to violate the law, for it shows a strong negative curvature (Figure 8-5). This apparent exception may be explained upon examination of the interlayer site coordination sphere. For large alkali cations, such as Rb and Cs, the coordination number is close to 12 oxygens, with all A–O distances similar. For small alkali cations, such as Na, the coordination number is closer to six, with six short Na–O bonds, and six much longer Na–O distances. Thus, in the mica interlayer position, small cations are effectively smaller than predicted by Vegard's Law because of the variation of coordination number.

(2) Excess volume of mixing

Vegard's Law works well for many end-member compounds of isomorphous series, but intermediate compositions within a solid solution series often display more complex X–V behaviour. X-ray data on unit-cell volume versus composition in solid solution series between end members A and B have been used to derive a general expression for the volume of A_xB_{1-x}:

$$V = xV_A + (1 - x)V_B + V_{excess} \tag{8-5}$$

where V_A and V_B are end-member volumes and V_{excess} is the excess volume of mixing. Nonzero volumes of mixing are commonly observed; thus, most crystal solid solutions are nonideal. There are no theoretical constraints on the sign or form of V_{excess}, and several formulations have been used.

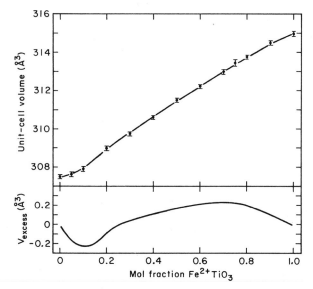

Figure 8-6. Excess volume of mixing for the solid solution series $MgTiO_3$–$FeTiO_3$ (from Newton and Wood, 1980; data of Bishop, 1976). Volume excess is negative near the small-cation end, and positive near the large-cation end

Excess volume of mixing is usually modelled by a second- or third-order polynomial (Thompson, 1967). Many solutions may be described using a symmetric excess function:

$$V_{excess} \propto x(1 - x) \tag{8-6}$$

or an asymmetric excess function:

$$V_{excess} \approx W_A x(1 - x)^2 + W_B x^2 (1 - x) \tag{8-7}$$

where W_A and W_B are constants.

Newton and Wood (1980) have demonstrated a more complex behaviour of the excess function in many common silicate mineral solutions. Careful unit-cell volume versus composition plots (Figure 8-6) reveal small but significant regions of *negative* excess volume near the small-cation end member in garnets, olivines, feldspars, pyroxenes, amphiboles, micas, ilmenites and other oxide and silicate mineral groups. A more familar positive and approximately parabolic excess function is observed nearer the large cation end member. Newton and Wood remark that "these excess function 'anomalies' must be related to crystal structure 'events' in the solid solution series." They postulate that a small concentration of large cations in smaller sites produces only 'local deformations,' so that volume of mixing is less initially than at greater concentration. Unfortunately, x-ray experiments, which yield a time- and space-averaged distribution of electron density, are not well-suited to resolve the nature of these local distortions.

C. Effects of composition on bond distance and polyhedral volume

An obvious effect of cation substitution into a polyhedron is to change the size of the polyhedron. Cation–anion distances in a disordered polyhedron will vary from unit cell to unit cell, depending upon the cation present in each specific location. X-ray diffraction techniques, however, can only be used to determine the mean bond distance, $\bar{d}$, averaged over the *entire* crystal for each symmetrically distinct polyhedron. The net change, $\Delta\bar{d}$, in mean bond distance is given by:

$$\Delta\bar{d} \approx \Delta X_2 \ (r_2 - r_1) \tag{8-8}$$

where ΔX_2 is the fractional change in occupancy of cation 2, and r_1 and r_2 are the ionic radii of the two cations. A consequence of equation (8-8) is that a specific substitution of one cation for another will have a predictable effect on polyhedral size, independent of structure. Consider, for example, substitution of octahedrally coordinated ferrous iron for magnesium, as illustrated for several compounds in Figure 8-7. Although the absolute values of Mg–O bond distances differ by a few per cent, the slopes of these several lines are equal. Note the similarity between this behaviour and the trends illustrated in Figures 6-4 and 7-2 for bond distance variation with temperature and pressure, respectively.

A similar and extremely useful plot may be constructed for the substitution of aluminum for tetrahedrally coordinated silicon. Figure 8-8 includes data on tetrahedra in framework silicates (feldspars), layer silicates (micas) and chain

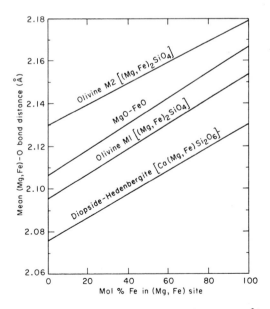

Figure 8-7. Bond distance versus mol percent Fe²⁺ in several compounds with octahedral Mg–Fe solid solution

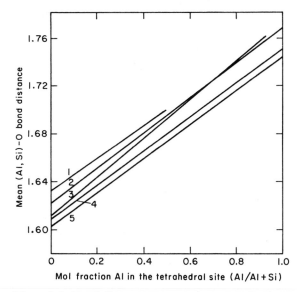

Figure 8-8. Bond distance versus mol percent aluminum in several groups of minerals with tetrahedral Al–Si solid solution. (1) $CaMgSi_2O_6$–$CaAl_2SiO_6$, (Hazen and Finger, 1977); (2) Layer silicates (Smith and Bailey, 1963), (3) Micas (Hazen and Burnham, 1973), (4) Feldspars (Smith, 1974), and (5) Framework silicates (Smith and Bailey, 1963)

silicates (pyroxenes), which display this common substitution. Again a set of parallel lines is observed. The utility of this relationship lies in the sensitivity of tetrahedral bond distances to Al–Si compositional variation. In a silicate with partially disordered tetrahedra, the similarity of silicon and aluminum electron density precludes direct determination of site composition by x-ray measurements. The site occupancies may be deduced, however, from the mean T–O bond distance.

Equations (8-1) and (8-8) may be combined, as implicit in the data of Figures 8-7 and 8-8, to relate the coefficient of compositional expansion to ionic radii of the substituting cations:

$$\gamma_l = \frac{1}{d} \left(\frac{\partial d}{\partial X} \right) \cong \frac{(r_2 - r_1)}{d} \tag{8-9}$$

In this relationship cation '1' may be a single ionic species, or a group of cations with:

$$r_i = \sum_{i=1}^{n} f_i r_i / \sum f_i \tag{8-10}$$

where f_i and r_i are the mole fraction and ionic radii of the ith cation, respectively.

The change in polyhedral volume, V_p, for a given compositional change is also a constant. This follows from equation (8-9), because $(r_2 - r_1)$ is generally small compared to d, and V_p is proportional to d^3. The change in polyhedral volume is thus given by the coefficient of volume compositional expansion:

$$\gamma_v \approx \frac{3(r_2 - r_1)}{d} \qquad (8\text{-}11)$$

This simple equation, like the analogous relationships for temperature coefficient, α (equation 6-18), and pressure coefficient, β (equation 7-14), may be used to predict the variation of structural parameters with composition.

It is probable that excess volumes of mixing parameters are commonly applicable to polyhedral volumes, just as they are to molar volume. Current crystal structure refinements, however, are not precise enough to detect small deviations from linearity in a composition-polyhedral volume plot. It is useful to remember, however, that equation (8-11) represents an approximation, which may be invalid if precision greater than a few per cent is required.

III. CONCLUSIONS

The principal structural response to a change in composition, within a given phase region, is a change in interatomic distances. Variation of unit-cell volume is approximately proportional to the cube of the mean ionic radius, as represented by Vegard's Law; however, excess volume of mixing commonly results in deviations of a few per cent from the ideal behaviour. Polyhedral volumes are also simply related to the differing ionic radii of substituting cations, as modelled by equations (8-9) and (8-11).

Changes in composition of a compound, as with changes in temperature or pressure, may be viewed as varying the ratio of polyhedral sizes, owing to the differing polyhedral coefficients of compositional expansion. Thus, for example, the substitution of ferrous iron for octahedral magnesium in silicates increases the ratio of octahedral-to-tetrahedral sizes, whereas the substitution of aluminum for tetrahedral silicon decreases this ratio. The importance of polyhedral size ratios in determining the stability of compounds will be explored in the next chapters.

REFERENCES

Bishop, F. C. (1976) *Partitioning of Fe^{2+} and Mg between ilmenite and some ferromagnesian silicates*, Ph.D. Thesis, Department of Geophysical Sciences, University of Chicago, Chicago, Illinois.

Carman, J. H. (1969) *The study of the system $NaAlSiO_4 - Mg_2SiO_4 - SiO_2 - H_2O$ from 200 to 5000 bars and 800°C to 1100°C and its petrologic applications*, Ph. D. Thesis, Department of Geosciences, Pennsylvania State University, University Park, Pennsylvania.

Eugster, H. P., and J. Munoz (1966) Ammonium micas: possible sources of atmospheric ammonium and nitrogen. *Science* **151**, 683–686.

Hazen, R. M., and C. W. Burnham (1973) The crystal structures of one-layer phlogopite and annite. *Am. Mineral.* **58**, 889–900.

Hazen, R. M., and L. W. Finger (1977) Crystal structure and compositional variation of Angra Dos Reis fassaite. *Earth Planet. Sci. Lett.* **35**, 357–362.

Hazen, R. M., and D. R. Wones (1972) The effect of cation substitutions on the physical properties of trioctahedral micas. *Am. Mineral.* **57**, 103–129.

Newton, R. C., and B. J. Wood (1980) Volume behavior of silicate solid solutions. *Am. Mineral.* **65**, 733–745.

O'Keeffe, M., and B. G. Hyde (1981) Nonbonded forces in crystals. In *Structure and Bonding in Crystals*, M. O'Keeffe and A. Navrotsky (eds), Academic Press, NY.

Pauling, L. (1960) *The Nature of the Chemical Bond*, Cornell University Press, Ithaca, NY, 3rd edition, 644 p.

Prewitt, C. T. (1977) Effect of pressure on ionic radii (abstr.). *Geol. Soc. Am. Abstracts with Programs* **9**, 1134.

Shannon, R. (1976) Revised effective ionic radii and systematic studies of interatomic distances in halides and chalcogenides. *Acta Crystallogr.* **A32**, 751–767.

Shannon, R. D., and C. T. Prewitt (1970) Revised values of effective ionic radii. *Acta Crystallogr.* **B26**, 1046–1048.

Slater, J. C. (1963) *Quantum Theory of Molecules and Solids*, McGraw-Hill, NY.

Smith, J. V. (1974) *Feldspar Minerals*, Springer-Verlag, NY.

Smith, J. V., and S. W. Bailey (1963) Second review of Al–O and Si–O tetrahedral distances. *Acta Crystallogr.* **16**, 801–811.

Thompson, J. B. (1967) Thermodynamic properties of simple solutions. In *Researches in Geochemistry*, Vol. **II**, P. H. Abelson (ed.), Wiley, NY, 340–361.

Vegard, L., and H. Dale (1928) Untersuchungen ueber Mischkristalle und Legierungen. *Zeits. Kristallogr.* **67**, 148–162.

Wones, D. R. (1963) Physical properties of synthetic biotites on the join phlogopite-annite. *Amer. Mineral.* **48**, 1300–1321.

Chapter 9

Continuous Structural Variations with Temperature, Pressure and Composition

CONTENTS

I   Introduction 177
II   The Structural Analogy of Temperature, Pressure and Composition . . . 178
    A   The Analogy 178
    B $T–P–X$ Surfaces of Constant Structure 179
    C   'Inverse Relationship' of Temperature and Pressure 180
    D   The 'Double' Internal X-ray Standard 186
III T, P and X variations of α, β and γ 187
    A $(\partial\alpha/\partial T)_{P,X}$ 187
    B $(\partial\alpha/\partial P)_{T,X} = -(\partial\beta/\partial T)_{P,X}$ 187
    C $(\partial\alpha/\partial X)_{T,P} = (\partial\gamma/\partial T)_{P,X}$ 188
    D $(\partial\beta/\partial P)_{T,X}$ 189
    E $(\partial\beta/\partial X)_{T,P} = -(\partial\gamma/\partial P)_{T,X}$ 190
    F $(\partial\gamma/\partial X)_{T,P}$ 191
    G   Summary 191
IV   Modelling Structural Variations—Distance Least Squares 192

I. INTRODUCTION

In the previous three chapters we have explored how crystal structures vary with temperature or pressure or composition. In this chapter we consider the structure variations that result from combinations of these changes in temperature–pressure–composition ($T–P–X$) space *within* a given phase region. In a $T–P–X$ phase region all structural changes are assumed to be continuous. (Discontinuous changes—phase transitions—are treated in Chapter Ten.) A long-term objective of high-temperature/high-pressure research is to provide structural $T–P–X$ equations of state for crystalline materials. Although such a goal is still distant for all but the simplest compounds, guidelines for structural equations of state may be developed by utilizing the close similarity between structural variations with temperature, with pressure and with composition.

II. THE STRUCTURAL ANALOGY OF TEMPERATURE, PRESSURE AND COMPOSITION

A. The analogy

Hazen (1977a) proposed that geometrical aspects of structure variation with temperature, pressure or composition are analogous in the following ways:

(1) The fundamental unit of structure for the purposes of the analogy is the cation coordination polyhedron. For a given type of cation polyhedron, a given change in temperature or pressure or composition has a constant effect on polyhedral size, regardless of the way in which polyhedra are linked. Polyhedral volume coefficients α_V, β_V and γ_V are thus independent of structure to a first approximation.

(2) Polyhedral volume changes with temperature or pressure or composition may be predicted from simple bonding parameters: cation–anion bond distance (d), cation coordination number (n), cation radius (r), formal cation and anion charge (z_c and z_a), and ionicity (S^2).

$$\alpha_V = \frac{1}{V}\left(\frac{\partial V}{\partial T}\right) \approx 12.0 \left(\frac{n}{S^2 z_c z_a}\right) \times 10^{-6\circ}\text{C}^{-1} \qquad (9\text{-}1)$$

$$\beta_V = \frac{-1}{V}\left(\frac{\partial V}{\partial P}\right) \approx 0.133 \left(\frac{d^3}{S^2 z_c z_a}\right) \times 10^{-6}\,\text{bar}^{-1} \qquad (9\text{-}2)$$

$$\gamma_V = \frac{1}{V}\left(\frac{\partial V}{\partial X}\right) \approx \frac{3(r_2 - r_1)}{d} \qquad (9\text{-}3)$$

as derived from equations (6-18), (7-14) and (8-11), respectively.

(3) As a corollary, in structures with more than one type of cation polyhedron, variations of temperature or pressure or composition all have the effect of changing the *ratios* of polyhedral sizes.

In these ways, therefore, changes in temperature, pressure, or composition may have similar effects on the geometry of a crystal structure. It is important to remember that the T–P–X analogy extends *only* to the geometrical aspects of a structure. Changes in temperature, pressure and composition produce very different effects on vibrational properties and electronic structure. Any aspect of the structure or physical properties of a solid that is primarily a function of electronic or thermal states will not display this behaviour. In practice, however, a great many compounds, including the majority of oxide and silicate minerals and mineral-like compounds, have structures that are not sensitive to changes in thermal or electronic parameters.

B. *T–P–X* surfaces of constant structure

All crystalline materials may be represented in *T–P–X* space by surfaces of constant molar volume (isochoric surfaces). One consequence of the structural analogy of temperature, pressure and composition is that for many substances isochoric surfaces are also surfaces of constant structure in *T–P–X* space (Hazen, 1977a). Consider, for example, the simple fixed structure of the solid solution between stoichiometric MgO and FeO. A single parameter, the unit-cell edge, completely defines the structure of this NaCl-type compound. Isochoric surfaces are constrained to be isostructural surfaces in *T–P–X* space (Figure 9-1), because variations in temperature or pressure or composition all change this parameter. Isochoric or isostructural surfaces may be approximately planar over a limited range of temperature, pressure and composition; however, α_V, β_V, and γ_V generally vary with T, P, and X, thus implying curved surfaces of constant volume (see Section III, below).

Isostructural surfaces exist for a large number of compounds. The magnesium–iron silicate spinel, $\gamma\text{-(Mg, Fe)}_2\text{SiO}_4$, is an example of special interest to geophysicists because of the mineral's supposed presence in the earth's mantle. In this cubic structure, which has only two variable parameters, the size of the silicon tetrahedral site is relatively constant with changes in temperature, pressure and Fe/Mg octahedral composition. Consequently, isostructural *T–P–X* surfaces for the octahedral component of the silicate spinels will also approximate planes of constant spinel structure. Note that the isostructural surfaces of (Mg, Fe)O and

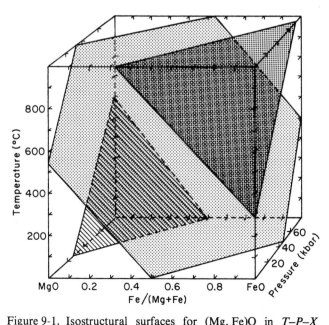

Figure 9-1. Isostructural surfaces for (Mg, Fe)O in *T–P–X* space

γ-$(Mg, Fe)_2SiO_4$ will be similar because both depend primarily on the size of the (Mg, Fe) octahedron.

All isostructural surfaces have certain features in common. Consider the slopes of such a surface:

$$\left(\frac{\partial P}{\partial T}\right)_{\diamond,X}, \left(\frac{\partial P}{\partial X}\right)_{\diamond,T} \text{ and } \left(\frac{\partial T}{\partial X}\right)_{\diamond,P}$$

where $\diamond$ designates partial differentials at constant structure (as well as constant molar volume), and positive ∂X is defined as substitution of a larger cation for a smaller one. It follows that:

$$\left(\frac{\partial P}{\partial T}\right)_{\diamond,X} > 0 \tag{9-4}$$

$$\left(\frac{\partial P}{\partial X}\right)_{\diamond,T} > 0 \tag{9-5}$$

$$\left(\frac{\partial T}{\partial X}\right)_{\diamond,P} < 0 \tag{9-6}$$

Even relatively complex structures, such as the biaxial alkali aluminosilicate, $(K, Na)AlSi_3O_8$, alkali feldspar, may have T–P–X surfaces of constant structure (see also Chapter Ten, Section II-B-5). Of course, not all compounds demonstrate this phenomenon. If a structure has more than two different types of cation polyhedra, for example, then a given change in T or P or X will commonly not be cancelled by any possible combination of changes of the other two, unless multiple chemical substitutions are invoked. Multiple compositional variables, of course, increase the dimensions of the T–P–X space under consideration.

C. 'Inverse relationship' of temperature and pressure

Geometrical structural changes, including changes in bond distances, bond angles and polyhedral distortions, occur as a result of changes in temperature and pressure. In numerous compounds these structural changes upon cooling from high temperature are similar to those upon compression. In other words, structural variations due to changes in temperature may be offset by variations due to changing pressure. When this inverse relationship applies it can be very useful in predicting structural changes with temperature and pressure. However, the relationship is *not* universal, and cannot be applied arbitrarily. The inverse relationship may obtain when:

(1) All polyhedra in a structure have similar ratios of expansivity to compressibility; i.e. α/β is a constant for all polyhedra of the structure; or,

(2) One polyhedron is relatively rigid (α and β are small) compared to the other polyhedra, which have similar α/β.

These conditions are fulfilled by numerous compounds including all materials with only one type of polyhedron, and a great many silicates with only one type of polyhedron other than the relatively rigid Si tetrahedra. From equations (9-1) and (9-2) for polyhedral α and β:

$$\alpha/\beta = 90n/d^3 \tag{9-7}$$

where n is coordination number and d is mean cation–anion bond distance. Thus, the 'inverse relationship' should obtain if n/d^3 is similar for all cation polyhedra in a structure. Coincidentally, several common cation polyhedra in rock-forming

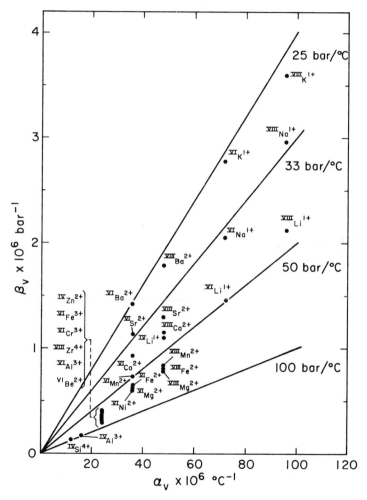

Figure 9-2. α versus β for oxygen-based polyhedra (after Hazen and Prewitt, 1977a). Values for α_V and β_V are those predicted from equations (9-1) and (9-2). Several lines of constant α/β are drawn. Cations are given in the form R_n^z, where n is the cation coordination number and z is the cation formal charge. (Reproduced by permission of the Mineralogical Society of America)

minerals have similar ratios of α to β; octahedral Mg, Fe^{2+}, Al and Fe^{3+} all have $\alpha/\beta \cong 65$ bar/°C. Thus, many minerals display the inverse relationship. Values of predicted α and β for many common cation polyhedra are illustrated in Figure 9-2.

The inverse relationship for temperature and pressure is rigorously true for simple fixed structures such as NaCl, cubic ZnS, CsCl or CaF_2. It is instructive, however, to examine in detail the extent to which this relationship is valid in structures with variable atomic positions.

Conformity with the inverse relationship is best visualized by plotting variable structural parameters versus molar volume or V/V_0 (Figure 9-3). If the inverse relationship is valid then all positional parameters in the structure, as well as bond distances and angles, distortion parameters and other variables, will have a continuous variation on such a plot.

For example, consider the silicate spinel γ-Ni_2SiO_4, which is a cubic compound with a single variable parameter. Figure 9-4 illustrates the inverse behaviour of this positional parameter with temperature and pressure. Such behaviour is expected because the silicon tetrahedron is relatively rigid with changes in temperature and pressure, whereas the Ni^{2+} octahedron expands or compresses in proportion to the molar volume.

The inverse relationship may obtain even in complex compounds, such as the alkali feldspar, $(K, Na)AlSi_3O_8$. In this common mineral a framework of aluminum and silicon tetrahedra (rigid polyhedra that undergo little change relative to the rest of the structure) are corner-linked and form large alkali cation sites. The lattice is triclinic giving several degrees of freedom for variation of the tetrahedral framework without changing symmetry. Hazen and Prewitt (1977b) determined high-pressure unit-cell parameters for the sodium feldspar (low albite) and compared these parameters with high-temperature values. All six triclinic unit-cell dimensions display the inverse relationship (Figure 9-5).

Another complex structure that displays the inverse relationship is the chain silicate $MnSiO_3$, pyroxmangite. This pyroxenoid has seven silicon tetrahedra in

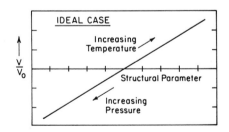

Figure 9-3. The ideal 'inverse relationship' for the variation of structural parameters with temperature and pressure. If the inverse relationship obtains then a continuous variation of the structural parameter versus V/V_0 will obtain

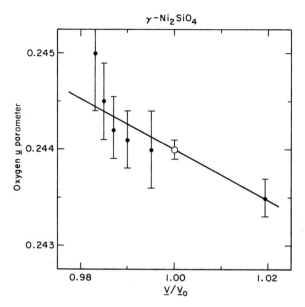

Figure 9-4. Silicate spinel u parameter versus V/V_0 at various temperatures and pressures. From data of Finger *et al.* (1978)

one repeat unit, with several crystallographically distinct 6-, 7- and 8-coordinated manganese polyhedra. High-pressure and high-temperature unit-cell data (Pinckney, personal communication) again show opposite trends of variation with temperature and with pressure (Figure 9-6).

Many other compounds studied at both elevated temperature and pressure display the inverse relationship. Corundum-type oxides (Finger and Hazen, 1980), olivines, $(Mg, Fe)_2SiO_4$ (Hazen, 1977b) and sanidine, $KAlSi_3O_8$ (Hazen, 1976) are among these other examples.

The pressure required to offset a 1°C increase in temperature is not the same for all compounds. In the case of MgO, 75 bar offsets 1°C. For Ni_2SiO_4, $NaAlSi_3O_8$ and $MnSiO_3$ the values of $(\partial P/\partial T)_\lozenge$ are 70, 20 and 40 bar/°C, respectively. The variability in $(\partial P/\partial T)_\lozenge$ is a result of the difference in α/β for Mg, Ni, Na and Mn polyhedra in these compounds. Note that in some cases $(\partial P/\partial T)_\lozenge$ is greater than the average geotherm of the earth's crust and upper mantle (25 bar/°C). Thus many minerals have *greater* molar volumes at depth in the earth than at the surface.

Silicon dioxide is one compound that should display the inverse relationship, yet Levien *et al.* (1980) concluded that the inverse relationship may not obtain in α-SiO_2 (quartz). However, the high-temperature data of Young (1962), which Levien *et al.* use to compare with their high-pressure data, are not sufficiently precise to demonstrate this conclusion. There are several structural features that are useful in discussing quartz variation with T and P; the polyhedral volume, the

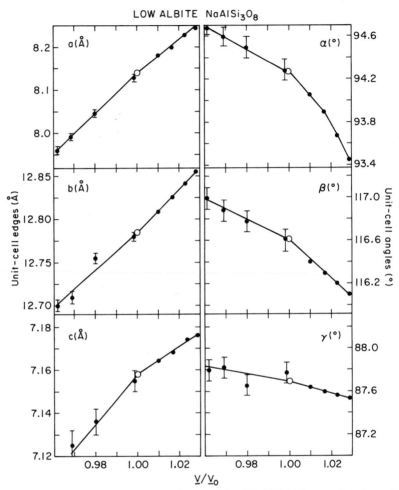

Figure 9-5. Unit-cell parameters of low albite ($NaAlSi_3O_8$) as a function of temperature and pressure. From data of Hazen and Prewitt (1977b)

polyhedral distortion, the axial ratio c/a, and the Si–O–Si angle between adjacent tetrahedra. The tetrahedral volume appears to be invariant with temperature, but decreases by approximately one standard deviation at 60 kbar. The tetrahedral distortions are unchanged with temperature, but increase by approximately one-half a standard deviation at 60 kbar. The axial ratio, c/a, increases with pressure and decreases with temperature. Another significant change in the structure is a large increase in Si–O–Si angle with temperature, and a correspondingly large decrease in the angle with pressure. Thus, except perhaps in details that are beyond the resolution of the present data, the inverse relationship holds for α-SiO_2.

Many compounds are not expected to display the inverse relationship. Some common polyhedra, such as those of Ca^{2+}, K^+ and Na^+ have much smaller values of α/β than polyhedra of Mg, Fe or Al. Thus silicates with a large alkali or alkaline earth cation plus Mg, Fe or Al will not display the inverse relationship in all structural parameters. Common examples of minerals that respond differently to temperature and pressure are diopside, $CaMgSi_2O_6$ (Levien and Prewitt, 1981), phlogopite mica, $KMg_3AlSi_3O_{10}(OH)_2$ (Hazen and Finger, 1978a), and grossular garnet, $Ca_3Al_2Si_3O_{12}$ (Hazen and Finger, 1978b). Even in these compounds, however, a qualitative inverse relationship may be observed, for the large Ca or K sites expand and compress more than the Mg or Al octahedral sites, which in turn expand or compress more than the tetrahedra of silicon.

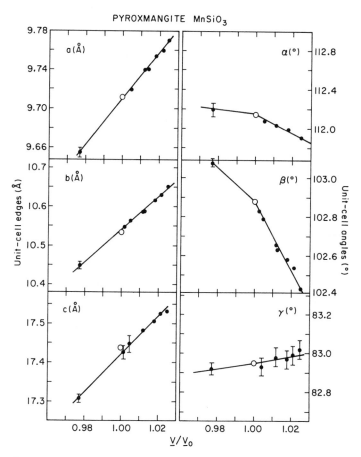

Figure 9-6. Unit-cell parameters of pyroxmanganite ($MnSiO_3$) versus V/V_0. Unit-cell parameters were measured at several temperatures and pressures (Pinckney, personal communication)

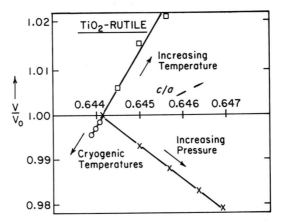

Figure 9-7. Axial ratio c/a for rutile (TiO_2) versus V/V_0. The c/a of rutile does not display the inverse relationship; the ratio increases both with temperature and with pressure. From data in Hazen and Finger (1981)

One compound examined at both high temperature and pressure deviates radically from the inverse relationship. Rutile-type TiO_2 is tetragonal with only one type of polyhedron ($^{VI}Ti^{4+}$). Rutile should follow the trend observed in other simple oxides, but a plot of c/a versus V/V_0 (Figure 9-7) shows a more complex behaviour: the axial ratio c/a increases *both* with temperature and with pressure. The inverse relationship would predict opposite behaviour with changes of T versus P. Surveys of several rutile-type compounds (Hazen and Finger, 1981; Rao, 1974) have revealed that although most isomorphs have increasing c/a with pressure, the temperature response is highly variable. Clearly the temperature variation of rutile-type compounds is not controlled by structure, because all of these compounds have the same polyhedral arrangements. Some other, and as yet unknown, property of the rutile-type dioxides and difluorides must be responsible for this anomalous behaviour.

D. The 'double' internal x-ray standard

The concept of a double x-ray standard, introduced in Chapter Four, section III-B-3, was based on the possibility of using two crystallographic directions which have very different α/β simultaneously for temperature and pressure calibration. The circumstances that might produce such a standard within a single material are evident from Figure 9-2. If one crystallographic direction is governed by a polyhedron with large α/β (e.g. Al or Mg in octahedral coordination) and another direction is governed by a polyhedron with small α/β (e.g. K or Ca) then an excellent calibration crystal might result. In the case of $CaCO_3$, calcite, the a-axis is controlled by C–O bonds, whereas the c-axis is controlled by Ca–O bonds, thus yielding an effective calibration for both temperature and pressure.

III. T, P AND X VARIATIONS OF α, β AND γ

Earlier in this chapter it was noted that the effects of temperature, pressure and composition on α, β and γ may be small, leading to near-planar isostructural T–P–X surfaces. The values of such quantities as $(\partial\alpha/\partial T)$ and $(\partial\beta/\partial T)$ are not zero, however, and the constraints on these and other derivatives are important in defining the equations of state of crystalline solids.

The temperature, pressure and composition derivatives of α, β and γ are of interest, yielding nine possible derivatives. Several of these, however, are equivalent:

$$-\left(\frac{\partial\beta}{\partial T}\right) = \frac{1}{V}\left(\frac{\partial V}{\partial P}\right)\left(\frac{1}{\partial T}\right)$$

$$= \frac{1}{V}\left(\frac{\partial V}{\partial T}\right)\left(\frac{1}{\partial P}\right) = \left(\frac{\partial\alpha}{\partial P}\right)$$

Similarly,

$$-\left(\frac{\partial\beta}{\partial X}\right) = \left(\frac{\partial\gamma}{\partial P}\right) \text{ and } \left(\frac{\partial\alpha}{\partial X}\right) = \left(\frac{\partial\gamma}{\partial T}\right)$$

Other derivatives are $(\partial\alpha/\partial T)$, $(\partial\beta/\partial P)$, and $(\partial\gamma/\partial X)$. These six nonequivalent terms are considered below.

A. $\left(\dfrac{\partial\alpha}{\partial T}\right)_{P,X}$

A plot of α versus temperature is shown in Figure 6-3. In general, for both unit-cell and polyhedral volumes, α is a monotonically increasing function versus temperature (though a small interval of negative α may be observed at very low temperature). As illustrated in Figure 6-3, $(\partial\alpha/\partial T)$ is positive and the α versus T plot levels off at high temperature.

When sufficient temperature–volume data are available it is possible to model thermal expansion as a polynomial function in T:

$$V = V_0 + aT + bT^2 + cT^3 \tag{9-8}$$

where a is the thermal expansion coefficient, and b is a positive coefficient related to the increase of α with temperature. The c coefficient may be either positive or negative, depending on the temperature range of T–V data. Near the Debye temperature c (as well as $\partial^2\alpha/\partial T^2$) is negative as α levels off to a near constant value. Unfortunately, there are insufficient T–V data for cation polyhedra to calculate these higher-order coefficients. The use of $\bar{\alpha}_{1000}$ throughout this monograph is thus a necessary, but oversimplifying, assumption.

B. $\left(\dfrac{\partial\alpha}{\partial P}\right)_{T,X} = -\left(\dfrac{\partial\beta}{\partial T}\right)_{P,X}$

The pressure derivative of thermal expansion, or temperature derivative of compressibility, has been measured in only a few crystals and no cation polyhedra.

These derivatives may be significant, however, invalidating the simple approximation that temperature and pressure variations of a structure are additive at high T and P. In the alkali halides $(\partial\alpha/\partial P)$ and $(\partial\beta/\partial T)$ were first studied experimentally by Bridgman (1940) who found that increased pressure decreases α (Figure 9-8). More recently Yagi (1978) measured the effects of temperature on compressibility of these alkali halides, and also found that compressibility increases with temperature (Figure 9-9). The bulk modulus of LiF, for example, was found to decrease from 665 kbar at room temperature to only 430 kbar at 800°C. CsCl has an even larger percentage reduction in bulk modulus, dropping from 182 to 104 kbar between 23° and 600°C. Graphs of bulk modulus versus T are presented in Figure 9-10. Note that Figures 9-8, 9-9 and 9-10 illustrate the same data in different ways.

These major volume second-derivative effects, unfortunately, are extremely difficult to measure because of uncertainties in simultaneous high-temperature and high-pressure calibration. It will thus be some time before precise $(\partial\alpha/\partial P)$ or $(\partial\beta/\partial T)$ are known for oxygen-based cation polyhedra. It may be assumed, however, that the sign of $(\partial\alpha/\partial P)$ is negative in most materials (and $(\partial\beta/\partial T) > 0$) because of the close relationship between bond rigidity, and α and β.

C. $$\left(\frac{\partial\alpha}{\partial X}\right)_{T,P} = \left(\frac{\partial\gamma}{\partial T}\right)_{P,X}$$

The coefficient of thermal expansion is essentially constant for differing cations of the same valence and coordination. For example, the replacement of Fe^{2+} for Mg^{2+} has little effect on thermal expansion. Thus, for simple cation substitution $(\partial\alpha/\partial X) \approx 0$.

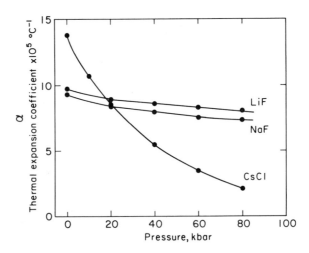

Figure 9-8. The variation of thermal expansion with pressure for three alkali halides. Thermal expansion decreases with pressure. (From data in Yagi, 1978)

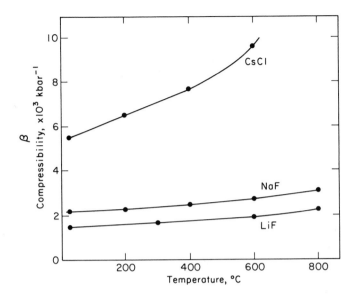

Figure 9-9. The variation of compressibility with temperature for three alkali halides. Compressibility increases with temperature. (From data in Yagi, 1978)

Coupled substitutions, such as ($^{VIII}Ca^{2+} + {^{IV}}Al^{3+} \rightleftharpoons {^{VIII}}K^{+} + {^{IV}}Si^{4+}$) or ($^{VI}Li^{+} + {^{VI}}Al^{3+} \rightleftharpoons 2{^{VI}}Mg^{2+}$) have a much more complex effect on structure and its expansion. In general, the addition of lower-valence cations to a structure will increase both polyhedral and bulk α because of the relatively greater expansion coefficients of cation polyhedra with low Pauling bond strength. Thus, the thermal expansion coefficient of $Na(AlSi_3)O_8$ is greater than that of $Ca(Al_2Si_2)O_8$.

D. $\left(\dfrac{\partial \beta}{\partial P}\right)_{T,X}$

In virtually all materials that do not undergo electronic transitions ($\partial\beta/\partial P$) is negative, as increased pressure decreases compressibility. This behaviour is intuitively reasonable, for a substance should become more rigid as it is compressed. From equation (9-2) it is evident that for a given type of polyhedron $\beta \propto V_p$ so that:

$$\left(\frac{\partial \beta}{\partial P}\right) \propto \left(\frac{\partial V}{\partial P}\right) = V\beta \propto V^2 \tag{9-9}$$

Thus, polyhedral ($\partial\beta/\partial P$) is approximately proportional to the square of the polyhedral volume. Consequently, large alkali sites undergo a larger *percentage* change in β versus pressure than small 3+ or 4+ sites.

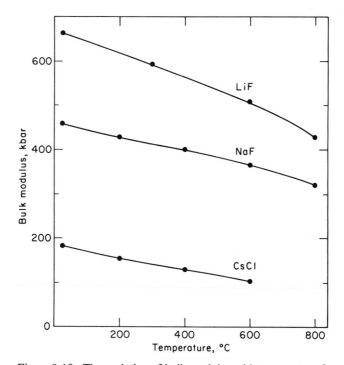

Figure 9-10. The variation of bulk modulus with temperature for three alkali halides. Bulk modulus decreases with temperature. (From data of Yagi, 1978)

The magnitude of these variations in β is not great in many oxygen-based materials. In magnesium oxide, MgO, linear compressibility decreases from about 2.0×10^{-4} to 1.7×10^{-4} $kbar^{-1}$ between 0 and 100 kbar (Decker *et al.*, 1972). In alkali halides, however, effects of pressure on β at 100 kbar are much greater. Compressibility of Na–Cl bonds in halite decreases from 1.4×10^{-3} at room pressure to 0.4×10^{-3} $kbar^{-1}$ at 100 kbar (Decker *et al.*, 1972). Highly compressible bonds thus show the greatest percentage variation in β with pressure.

Variation of polyhedral compressibility with pressure has yet to be demonstrated in complex oxides or silicates. It is likely that careful high-pressure structure studies to 100 kbar of a sodium or potassium silicate such as $NaAlSi_2O_6$ (jadeite) or $KAlSi_3O_8$ (sanidine) would reveal a decrease in alkali–oxygen bond compression at the highest pressures.

E. $\left(\dfrac{\partial \beta}{\partial X} \right)_{T,P} = - \left(\dfrac{\partial \gamma}{\partial P} \right)_{T,X}$

The effects of compositional change on compressibility are implicit in the bulk modulus–volume relationship. Substituting an amount ΔX of a larger cation, or one with lower valence, increases polyhedral compressibility by an amount defined by:

$$\Delta\beta = \frac{7.5\Delta X}{S^2 z_a} \left(\frac{d_2^3}{z_{c2}} - \frac{d_1^3}{z_{c1}} \right)$$ (9-10)

Thus, $(\partial\beta/\partial X) > 0$, assuming positive X is substitution of a larger cation.

F. $\left(\dfrac{\partial\gamma}{\partial X} \right)_{T,P}$

To a first approximation γ is linear with respect to composition for a given polyhedron [equation (8-11)]. Thus, $(\partial\gamma/\partial X) \approx 0$. However, in terms of bulk properties, and perhaps polyhedral properties as well, excess volumes of mixing may be real and nontrivial in functional form (see Chapter Eight, section II-B-2). Detailed studies of polyhedral volume versus site composition will be required to detect these small effects. Until that time it must be assumed that $(\partial\gamma/\partial X) \approx 0$.

G. Summary

The following generalizations may be made regarding the temperature, pressure and composition derivatives of α, β and γ:

$$\left(\frac{\partial\alpha}{\partial T} \right)_{P,X} > 0$$ (9-11)

$$\left(\frac{\partial\alpha}{\partial P} \right)_{T,X} = - \left(\frac{\partial\beta}{\partial T} \right)_{P,X} < 0$$ (9-12)

$$\left(\frac{\partial\alpha}{\partial X} \right)_{T,P} = \left(\frac{\partial\gamma}{\partial T} \right)_{P,X} \approx 0$$ (9-13)

$$\left(\frac{\partial\beta}{\partial P} \right)_{T,X} < 0$$ (9-14)

$$\left(\frac{\partial\beta}{\partial X} \right)_{T,P} = - \left(\frac{\partial\gamma}{\partial P} \right)_{T,X} > 0$$ (9-15)

$$\left(\frac{\partial\gamma}{\partial X} \right)_{T,P} \approx 0$$ (9-16)

In the cases of equations (9-12) and (9-14) especially, but also equations (9-11) and (9-15), these derivatives of α, β and γ cannot be ignored in the extrapolation of structures to elevated temperature and pressure. The simplified model presented in section II of this chapter is valid only for low temperature and pressure, therefore, and much work at combined temperature and pressure is needed. The development of accurate pressure and temperature calibration in such experiments will remain for some time the principal obstacle in efforts to determine structural variations in T–P–X space.

IV. MODELLING STRUCTURAL VARIATIONS–DISTANCE LEAST SQUARES

In the previous chapters we have established empirical methods for determining the variation of average cation–anion bond distances with temperature, pressure and composition. It has been stated that changes in a structure are not always determined by polyhedral sizes alone, however; bond angle variations are also important. Even if these angular changes are not known, it may be possible to model structures as a function of temperature, pressure and composition using distance least-squares (d.l.s.) techniques.

The d.l.s. procedure is dependent on the fact that the number of structural variables (unit-cell and atomic positional parameters) is generally less than the number of symmetrically independent cation–anion bond distances. If the symmetry, topology and expected values for bond distances are used as input, then a least squares fit of distance may be very successful in modelling details of the structure.

The distance least squares method, first developed by Meier and Villiger (1969), has been employed primarily to determine the extent to which bond distances alone can be used to understand details of atomic arrangements (Baur, 1972; Dempsey and Strens, 1976). In the case of unknown structures, d.l.s. may be employed to generate initial coordinates for refinement (Dollase and Baur, 1976). In the present context, however, the d.l.s. technique is of greatest interest in the prediction of crystal structures at high temperature and pressure using the empirical values of bond distances derived from equations (9-1), (9-2) and (9-3).

Dempsey and Strens (1976) were probably the first to apply d.l.s. to the calculation of thermal expansion and compressibility. They were successful in matching the observed α and β for Mg_2SiO_4 and Fe_2SiO_4 (the olivines forsterite and fayalite) using bond expansivities and compressibilities from simple oxides. Stolper (1977, unpublished manuscript 'Distance least squares modelling of crystal structures') has subsequently used d.l.s. to compare observed and calculated temperature variations of the cristobalite (SiO_2) structure, and the T–P variations of the olivine and spinel modifications of $(Mg, Fe)_2SiO_4$. Stolper demonstrated that d.l.s. could be successfully employed to model many aspects of changing structure with temperature and pressure, but that some parameters such as T–O–T angles in low cristobalite or polyhedral distortions in olivines may not be well modelled by this procedure. Several studies now under way, notably at Harvard University (Bish and Burnham, 1980), will help to determine the success of d.l.s. methods in predicting structural details of complex crystals at nonambient conditions.

REFERENCES

Baur, W. (1972) Computer-simulated crystal structures of observed and hypothetical Mg_2SiO_4 polymorphs of low and high density. *Am. Mineral.* **57**, 709–731.

Bish, D. L., and C. W. Burnham (1980) Structure energetics of cation-ordering in orthopyroxene and Ca-clinopyroxenes. *Geol. Soc. Am. Abstracts with Programs* **12**, 388.

Bridgman, P. (1940) the compression of 46 substances to 50,000 kg/cm². *Proc. Am. Acad. Arts. Sci.* **74**, 21–51.

Decker, D. L., W. A. Bassett, L. Merrill, H. T. Hall, and J. D. Barnett (1972) High-pressure calibration; a critical review. *J. Phys. Chem. Reference Data* **1**, 773–836.

Dempsey, M., and R. G. J. Strens (1976) Modelling crystal structures. In *The Physics and Chemistry of Minerals and Rocks* R. G. J. Strens (ed.), John Wiley and Sons, NY, 443–458.

Dollase, W. A., and W. H. Baur (1976) The superstructure of meteoritic low tridymite solved by computer simulation. *Am. Mineral.* **61**, 971–978.

Finger, L. W., and R. M. Hazen (1980) Crystal structure and isothermal compression of Fe_2O_3, Cr_2O_3, and V_2O_3 to 50 kbar. *J. Appl. Phys.* **51**, 5362–5367.

Finger, L. W., R. M. Hazen and T. Yagi (1978) Crystal structures and electron densities of nickel and iron silicate spinels at elevated temperature or pressure. *Am. Mineral.* **64**, 1002–1009.

Hazen, R. M. (1976) Sanidine: predicted and observed monoclinic-to-triclinic reversible transformations at high pressure. *Science* **194**, 105–107.

Hazen, R. M. (1977a) Temperature, pressure and composition: structurally analogous variables. *Phys. Chem. Minerals* **1**, 83–94.

Hazen, R. M. (1977b) Effects of temperature and pressure on the crystal structure of ferromagnesian olivine. *Am. Mineral.* **62**, 286–295.

Hazen, R. M., and L. W. Finger (1978a) The crystal structures and compressibilities of layer minerals at high pressures. II. phlogopite and chlorite. *Am. Mineral.* **63**, 293–296.

Hazen, R. M., and L. W. Finger (1978b) Crystal structures and compressibilities of pyrope and grossular to 60 kbar. *Am. Mineral.* **63**, 297–303.

Hazen, R. M., and L. W. Finger (1981) Bulk moduli and high-pressure crystal structures of rutile-type compounds. *J. Phys. Chem. Solids* **42**, 143–151.

Hazen, R. M., and C. T. Prewitt (1977a) Effects of temperature on interatomic distances in oxygen-based minerals. *Am. Mineral.* **62**, 309–315.

Hazen, R. M., and C. T. Prewitt (1977b) Linear compressibilities of low albite: high-pressure structural implications. *Am. Mineral.* **62**, 554–558.

Levien, L., C. T. Prewitt and D. J. Weidner (1980) Structure and elastic properties of quartz at pressure. *Am. Mineral.* **65**, 920–930.

Levien, L., and C. T. Prewitt (1981) High-pressure structural study of diopside. *Am. Mineral.* **66**, 315–323.

Meier, W. M., and H. Villiger (1969) Die Methode der Abstandsverfeinerung zur Bestimmung der Atomkoordinaten idealisierter Gerüststrukturen. *Z. Krist.* **129**, 411–423.

Rao, K. V. K. (1974) Thermal expansion and crystal structure. *Am. Inst. Physics Conference Proc.* **17**, 219–230.

Yagi, T. (1978) Experimental determination of thermal expansivity of several alkali halides at high pressures. *J. Phys. Chem. Solids* **39**, 563–571.

Young, R. A. (1962) Mechanism of phase transition in quartz. *AFOSR-2569 (Final Report, Project No. A-447) Engineering Experimental Station, Georgia Institute of Technology.*

Chapter 10

Structural Variations and the Prediction of Phase Equilibria

CONTENTS

I	Topological Classification of Phase Transitions	. 194
II	Geometrical Limits to Phase Stability	. 196
	A Radius Ratio Limits	. 196
	B Reversible Transitions—Polyhedral Tilting	. 197
	(1) Geometrical aspects of tilt transformations	. 197
	(2) Symmetry of tilt transformations	. 199
	(3) Preservation of single crystals	. 200
	(4) Twinning	. 200
	(5) Examples of polyhedral tilt transitions	. 201
	C Reconstructive Transitions—Layer Silicates	. 203
	(1) The talc-like layer	. 203
	(2) The serpentine layer	. 207
	D Martensitic Transitions—Chain Silicates	. 210
III	Polyhedral Stability Fields	. 211
	A Critical Limits to Bond Distances	. 211
	B Limits to Polyhedral Stability	. 213
IV	Conclusions	. 214

I. TOPOLOGICAL CLASSIFICATION OF PHASE TRANSITIONS

One of the great challenges of solid-state science is the prediction of solid–solid phase equilibria. Although thousands of phase transitions have been documented experimentally, little success has been achieved in calculating the stability of crystalline matter from either theoretical or empirical models. It is possible, however, using concepts developed in the previous chapters, to predict the stability of certain compounds that possess geometrically limited structures. Phase transitions may occur in these compounds when adjacent structural elements (e.g. polyhedra) reach critical size limits by continuous variations with temperature, pressure or composition.

In order to facilitate a geometrical approach to phase transformations it is

194

useful to review the topological classification scheme for phase transitions of Buerger (1951, 1972) as elaborated by Megaw (1973). The topology of a structure is defined in Buerger's scheme as the set of linkages or bonds between all nearest neighbour atoms. In an ionic compound, for example, all cation–anion bonds define the topology. Phase transitions are classified according to the extent to which topology is altered. (Changes in order–disorder parameters, including order–disorder transitions, are not considered in the context of this monograph.)

Reversible or displacive phase transformations involve no significant alteration of coordination polyhedra or their linkages, except perhaps for subtle shifts in the coordination of large cation sites. Reversible transformations are usually characterized by a change in symmetry, but transitions involving small discontinuities in atomic positions with no symmetry change may also occur (Hazen and Finger, 1979). Displacive transitions have relatively small activation energies, take place rapidly and reversibly, and single crystals are undamaged through the transition. A significant property of compounds that undergo displacive transformations is that the phases cannot be preserved metastably. High-temperature and high-pressure crystallographic studies are thus mandatory.

Megaw (1973) subdivided reversible phase transitions into groups based on the mechanism of transformation. Several of these types (hydrogen-hopping, orientation-switching and electronic transitions) are dependent upon thermal or electronic characteristics, and the transitions are thus not directly dependent on geometrical aspects of the structure. On the other hand, 'pure displacive transitions', which are characterized by small shifts in atomic positions, may often be described in terms of geometrical limitations. For example, some pure displacive transitions take place as the result of an extension or collapse of a structure about a large cavity or alkali cation coordination polyhedron, which varies with temperature, pressure or composition. Phase equilibria may be modelled if changes in large-site volume in $T–P–X$ space are known.

The other extreme type of phase transformation is the reconstructive transition, in which many bonds are broken, thus leading to a completely new topology. Groups of atoms as large as several polyhedra may remain intact in some reconstructive phase changes, but in others the great majority of bonds are broken and re-formed. Many of these transitions have been discussed in terms of limiting geometries, such as cation or polyhedral size ratios. Predictions of phase equilibria based on calculated $T–P–X$ structural variations are thus possible.

Martensitic or shear transformations are phase changes intermediate between reversible and reconstructive. Large portions of the structure are unchanged, but layers or slabs of structure are displacd with respect to each other. Bond topology is affected along slab boundaries, but the majority of bonds may remain intact. Typical examples of martensitic transitions in ionic compounds include NaCl-type to CsCl-type, rutile-type to $\alpha-PbO_2$-type, and some pyroxene to pyroxenoid transitions. Martensitic transitions may be rapid, and in some instances single crystals may be preserved, although atomic topology is significantly altered.

The topological classification of phase transitions has proved useful in describing energetics and mechanisms of transformations, although in practice all gradations between reversible, martensitic and reconstructive transitions are possible. In the next section examples of these three types of phase transitions, which may be geometrically limited, are reviewed.

II. GEOMETRICAL LIMITS TO PHASE STABILITY

Phase transitions of geometrically limited structures arise from the misfit of adjacent structural components. In many cases these misfits can be quantified in terms of the relative or absolute sizes of the structural units, and $T–P–X$ limits to stability may be equated to those temperatures, pressures and compositions at which the size restrictions are violated. Using the relationships summarized in Chapter Nine it is possible to predict changes in absolute and relative sizes of cation–anion bonds of a crystal structure as a function of temperature, pressure and composition. Thus, in the case of geometrically controlled stability, our knowledge of continuous structural variation leads to predictions of discontinuous structural changes. In the subsequent sections a variety of geometrically limited structures are considered in detail.

A.   Radius ratio limits

Perhaps the first suggestion of a geometrical limit to phase stability was Pauling's (1939) discussion of radius ratios. Pauling noted that carbonates of the form $R^{2+}CO_3$ assumed the calcite structure if R^{2+} had a radius smaller than 1.10 A, but crystallized in the aragonite structure for larger cations. He then proposed that the relative stability of these carbonates was a function of the cation/anion radius ratio, ρ. Similar radius ratio limits were proposed for a variety of simple compounds. For example, compounds of the form $M^{2+}X_2^-$ assume the rutile structure (M in 6-coordination) if $\rho < 0.73$, or the fluorite structure (M in 8-coordination) if $\rho > 0.73$. In these structures larger cations are more stable in the higher-coordinated polyhedra; that is, large ρ favours higher coordination number.

Temperature, pressure and composition all influence ionic radius ratios, and thus have analogous effects on the stabilities of simple ionic structures. An increase in pressure will stabilize the denser, closepacked structures, increasing ρ. An increase in temperature stabilizes the less dense, open structures, thus decreasing ρ. Substitution of different cations will also change ρ. If the stability of a given ionic structure is a function of cation/anion radius ratio then the $T–P–X$ transition surface for that substance may also be a surface of constant ρ. Such a surface will not, in general, be isostructural because the absolute values of cation and anion radii are not fixed.

Several generalizations may be made about the $T–P–X$ transition surfaces that are constrained by constant ionic radius ratio. If ρ is constant along a transition surface, then $(\partial P/\partial T)_{\rho,X}$ will always be positive, because increased pressure may be offset by increased temperature. If the substituting cation is larger, then $(\partial P/\partial X)_{\rho,T}$ will be negative and $(\partial T/\partial X)_{\rho,P}$ positive, whereas if the substituting cation is smaller, then $(\partial P/\partial X) > 0$ and $(\partial T/\partial X) < 0$. In other words, for simple ionic structures, an increase in temperature, a decrease in pressure, or substitution of a smaller cation all have the same effects on structure: all three changes decrease ρ.

B. Reversible transitions—polyhedral tilting*

Polyhedral tilting is a common pure displacive phase transition mechanism in ionic materials. Polyhedral tilt transitions may be recognized by five criteria:

1. Polyhedral tilt transitions occur in ionic compounds with corner-linked, rigid cation–anion groups, such as tetrahedral Al–Si groups in framework silicates or octahedral groups in perovskites. The framework of rigid polyhedra may form large cation sites, such as Na^+ in $NaAlSi_3O_8$ (albite) or Ba^{2+} in $BaFeSi_4O_{10}$ (gillespite), or cavities or channels, as in SiO_2 (quartz) or zeolites. Polyhedral tilt transitions occur when polyhedral elements of the rigid framework tilt owing to the changing size of the large cation site or cavity with changing pressure or temperature.
2. The transition is between a high-symmetry or less distorted form (stable at higher temperature or lower pressure) and a low-symmetry or more distorted form (stable at lower temperature or higher pressure).
3. The transition is rapid, reversible, and nonquenchable; single crystals are preserved through the phase change.
4. Twinning is commonly introduced in transforming from the higher symmetry to the lower symmetry form. The twin law is a symmetry operator of the high-symmetry phase that is lost in the transition.
5. The transition is geometrically related to the size of the large site or cavities; therefore, $(dP/dT) = (\partial P/\partial T)_v$, of large site = the ratio of thermal expansivity to compressibility (α/β) for the large site.

Of the five characteristics common to all polyhedral tilt transitions, (3) and (4) are generally true of pure displacive transitions. In some cation displacement transitions (a type of pure displacive transition in which cations shift off symmetry positions in polyhedra) characteristic (1), as in $BaTiO_3$, or characteristic (2), as in $CaCO_3I \rightleftharpoons II$, may also be observed. We know of no cation displacement transition, however, in which all five characteristics obtain. In the $BaTiO_3$ transition dP/dT is negative (Clarke and Benguigui, 1977), and in $CaCO_3I \rightleftharpoons II$ the structure is not composed of corner-linked polyhedra (Merrill and Bassett, 1975). Many examples of polyhedral tilt transitions are known, and several are listed in Table 10-1.

(1) Geometrical aspects of tilt transformations

All polyhedral tilt transitions result from the greater compressibilities and thermal expansivities of weakly bonded large cation polyhedra compared to the

---

* This section is derived from Hazen and Finger (1979).

Table 10-1. Polyhedral tilt transitions

Formula	Mineral	Transition: T(°C)	P(kbar)	dP/dT (bar/°C)	Space Group high form	low form	Reference
SiO_2	Quartz	573	0	+40	$\left.\begin{array}{l}P6_222\\P6_422\end{array}\right\}$	$P3_221$ $P3_121$	Cohen et al. (1974)
$NaNbO_3$	Perovskite-type						
	$a^\circ a^\circ a^\circ \leftrightarrow a^\circ a^\circ c^+$	640	0	—	$Pm3m/C$	$C4/mmb$	Megaw (1974)
	$a^\circ a^\circ c^+ \leftrightarrow a^- b^\circ c^+$	575	0	—	$\left.\begin{array}{l}C4/mmb\\Ccmm\end{array}\right\}$	$Ccmm$ $Pnmm$	
	$a^- b^\circ c^+ \leftrightarrow a^- b + c^+$	520	0	—			
$MgSiO_3$	Clinoenstatite	980	0	—	$C2/c$	$P2_1/c$	Smyth (1974)
$NaAlSiO_4$	Carnegieite	707	0	+125			Cohen and Klement (1976)
$KAlSi_3O_8$	Sanidine	$\left.\begin{array}{ll}23 & 24\end{array}\right.$		+15	$C2/m$	$C\bar{1}$	Hazen (1976)
$NaAlSi_3O_8$	High Albite/Monalbite	~1100	0				
$NaAlSi_2O_6 \cdot H_2O$	Analcite						
	I ↔ II	23	7	+57	Tetragonal	Monoclinic	Hazen and Finger (1979)
	II ↔ III	23	19	—	Monoclinic	Triclinic	
$BaFeSi_4O_{10}$	Gillespite	23	12	—	$P4/ncc$	$P4_22_12$	Hazen (1977a)
$Al_3Mg_2(Si_5Al)O_{18}$	Cordierite (disordered)	>1400	0	—	$P6/mcc$	$Cccm$	Langer and Schreyer (1969)
$Fe_{12}^{2+}Fe_6^{3+}Si_{12}O_{40}(OH)_{10}$	Deerite	~600		—	$Pna2$ or $Pnma$	$P2_1/a$	Agrell (pers. comm.)

surrounding corner-linked, rigid polyhedra. As a high-symmetry or less-distorted structure (the 'high' form) is cooled or compressed toward a polyhedral tilt transition, the large site or cavity becomes smaller at a greater rate than the rigid, linked polyhedra. A transition to a 'low' form occurs at a critical minimum size of the large site. In general, a slight structural rearrangement by tilting of corner-linked, small-cation polyhedra causes a decrease in large-site volume with little or no change in the size of rigid polyhedra. Tilting of corner-linked polyhedra will have an effect on lattice parameters that is dependent on structure type and degree of tilt (Megaw, 1973). Each structure type, therefore, must be analysed separately in relating structural changes to unit-cell dimensions.

Polyhedral tilt transitions may be described geometrically, because the size of the large site or cavity governs which form is observed. The high form has the greater large-site volume, and this volume can be reduced to the critical size by a reduction in temperature, an increase in pressure or substitution of a smaller cation into the large site. (Note that Buerger and Megaw treat these transitions *only* in terms of temperature variation.) The nature of the transition in $T-P-X$ space, therefore, will be similar for all polyhedral tilt transitions, as illustrated in Figure 10-1. Note, for example, that dP/dT will always be positive, because a lowering of pressure has the same effect on large-site volume as a raising of temperature. Furthermore, the magnitude of dP/dT, a macroscopic property, will be similar to the large site polyhedral $\bar{\alpha}/\bar{\beta}$ (Chapter Nine, equation 9-7). Polyhedral tilt transitions, therefore, are examples of the analogy of geometrical structure variations due to changes in temperature, pressure and composition.

(2) Symmetry of tilt transformations

The tilted form is the lower symmetry or more distorted form in polyhedral tilt transitions. Strens (1967) demonstrated that in reversible transitions involving a change of symmetry, the high-temperature form is also the form with greater degeneracy of normal vibration modes and consequently has the higher symmetry. It follows that the low-pressure form will also have the higher symmetry because dP/dT is positive. Strens also emphasized the important point that reversible transitions in which there is a discontinuity of volume need not be accompanied by a change in symmetry.

A symmetry change in polyhedral tilt transitions generally involves the removal of a symmetry operator of the high-symmetry form. In second-order structural transitions (i.e. transitions in which there is no discontinuity in volume, entropy or enthalpy) this relationship is rigorously true (Boccara, 1968), because the lower symmetry form must have a unit cell and space group that are subgroups of the higher symmetry form. Subgroup–supergroup relationships are seen, for example, in the $\alpha \rightleftharpoons \beta$ quartz, monalbite $\rightleftharpoons$ high albite, and several perovskite transitions. In first-order polyhedral tilt transitions (e.g. those with nonzero ΔV) this space-group relationship is not always observed, as in the gillespite I $\rightleftharpoons$ II phase transition (Hazen, 1977a).

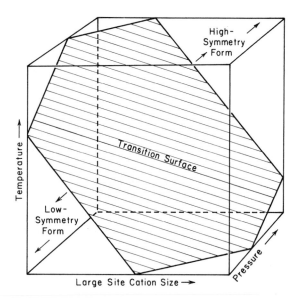

Figure 10-1. Phase transition T–P–X surface for a polyhedral tilt transition. (From Hazen and Finger, 1979, reproduced by permission of Gordon and Breach Science Publishers)

(3) Preservation of single crystals

An important characteristic of polyhedral tilt transitions, at least from an experimentalist's point of view, is that crystals are preserved through the transition. The slight rearrangements of structural elements, even if several weak metal–oxygen bonds are broken, are not sufficient to destroy the external morphology of the sample. It is possible, therefore, to measure the variation of directional properties in crystals through the transition.

Polyhedral tilt transitions are nonquenchable. A high-pressure or high-temperature form will invert to the room-condition structure immediately upon lowering of temperature or pressure. In some first-order polyhedral tilt transitions (e.g. gillespite I $\rightleftharpoons$ II) a slight hysteresis may be present, but this effect is small. These transitions, therefore, must be studied *in situ* at high temperature or high pressure.

(4) Twinning

Twinning is a common, if not ubiquitous, consequence of a symmetry reduction in polyhedral tilt transitions. The topologies of both low- and high-symmetry forms are closely related, if not identical, and there exists at least a two-fold ambiguity in the orientation of the low-symmetry form because of the removal of a symmetry element. Twinning is introduced, therefore, in transitions from high- to low-symmetry forms, and the twin law is a symmetry element lost during transformation from the high-symmetry phase.

Polyhedral tilt twinning may reveal useful information if the symmetry relations of the transition twins are known. The presence of this type of twinning in a crystal may provide evidence that the crystal at some time existed in the high-symmetry form and was subsequently cooled below the transition. In quartz, for example, Dauphine twins indicate cooling from the untwinned β phase; twinning in cordierite, deerite, nepheline and other collapsed mineral structures may also provide information on geothermal history. The absence of the specific type of twinning associated with polyhedral tilting, on the other hand, may imply a low temperature of crystallization within the low-symmetry phase region. In addition, twinning induced in high-pressure, collapsed forms, such as gillespite II, may aid in the interpretation of the structure of the high-pressure form (Hazen, 1977a) as identification of the twin law should indicate the nature of the large site distortion. Note, however, that compounds with polyhedral tilt transitions at high pressure will not show the transition-associated twinning under room conditions.

Polyhedral tilt twins complicate the determination of some structures under room conditions. One solution to this problem is to determine the high-symmetry structure at high temperature first and then proceed to determine the related structure of the twinned modification under room conditions. It should also be noted that synthetic crystals grown below the transition temperature are less likely to be twinned and may provide better material for single-crystal studies or commercial use.

(5) Examples of polyhedral tilt transitions

Sanidine The alkali feldspar sanidine (K, Na)AlSi$_3$O$_8$, with disordered aluminum and silicon in tetrahedral sites, exists in both monoclinic and triclinic forms depending on the temperature, pressure and composition. The two forms (Figure 10-2) are topologically equivalent, but the triclinic structure has a more distorted framework owing to collapse about the alkali site. The transition between monoclinic and triclinic varieties is a good example of a polyhedral tilt transition. The transition surface, which has been determined experimentally only in the T–X and P–X planes, is illustrated in Figure 10-3. Note that because the tetrahedra vary little with changes in T, P, or alkali X, that the transition surface illustrated in Figure 10-3 should also be an isostructural surface. This is borne out by the fact that unit-cell dimensions of feldspars at transition in the T–X and P–X planes have the same unit-cell dimensions (Hazen, 1977b). Furthermore, dP/dT of the transition (15 bar/°C) is similar to α/β for an 8-coordinated (Na, K) site.

Analcite Analcite is a zeolite mineral with ideal formula NaAlSi$_2$O$_6$·H$_2$O. The structure is a framework of Al and Si tetrahedra, which define cavities and channels for Na and H$_2$O. Under room conditions most analcites are pseudo-cubic,

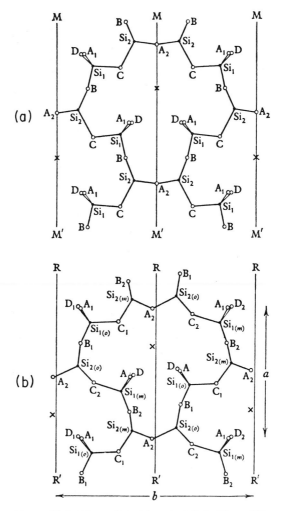

Figure 10-2. Monoclinic (a) and triclinic (b) modifications of the alkali feldspars $(K,Na)AlSi_3O_8$. (From Deer *et al.*, 1963, reproduced by permission of Longman's Group)

but several lower-symmetry varieties exist because of Al, Si ordering between tetrahedra. Hazen and Finger (1979) studied the lattice constants of analcite and discovered at least three reversible transitions to lower symmetry, twinned forms in analcite at high pressure (Figure 10-4). Differential thermal analysis of one of these transitions (at ≈ 8 kbar) revealed a slope $dP/dT \approx 57$ bar/°C (Rosenhauer and Mao, 1975). The analcite phase transitions thus fulfill all criteria of polyhedral tilting.

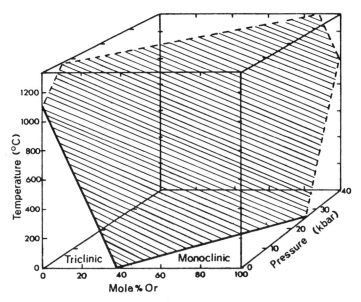

Figure 10-3. Polyhedral tilt transition surface in $T–P–X$ space for the monoclinic-to-triclinic transition in alkali feldspars. (From Hazen, 1976, *Science* **194**, 105–107. Copyright 1976 by the American Association for the Advancement of Science)

C. Reconstructive transitions—layer silicates

Some compounds with extensive edge- or face-sharing of polyhedra have stability limits based on the misfit of adjacent structural units in regions of $T–P–X$ space. In these compounds bond bending (as observed in polyhedral tilting) is severely restricted, and reconstructive transitions should occur at geometrical limits. Even if geometrical limits are well known it is not generally possible to predict phase equilibria; structural limits can only be used to map regions of $T–P–X$ space in which a structure *cannot* occur. Even so, it is of more than academic interest to predict the possible existence of a phase at a given temperature, pressure and composition.

(1) The talc-like layer*

The talc-like layer (Figure 8-2), is composed of an octahedral layer usually of Mg and Fe^{2+}, sandwiched between tetrahedral (Al, Si) layers. This three-layer unit is a basic building block of many silicates, including talc, $(Mg, Fe)_6 Si_8 O_{20}(OH)_4$; mica, $K(Mg, Fe)_3 AlSi_3 O_{10}(OH)_2$; and chlorite, $(Mg, Al, Fe)_{12}(Si, Al)_8 O_{20}(OH)_{16}$.

---

* This section is derived from Hazen and Wones (1978).

204

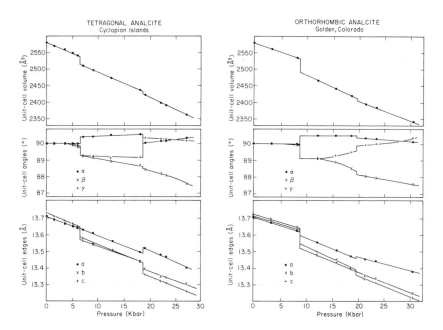

Figure 10-4. Unit-cell dimensions versus pressure for the pseudo-cubic zeolite mineral, analcite ($NaAlSi_2O_6 \cdot H_2O$). $4A$ is a tetragonal analcite from the Cyclopian Islands, and $4B$ is an orthorhombic analcite from Golden, Colorado. Both specimens have two volume discontinuities (at about 7.5 and 19 kbar) as well as changes in dimensional symmetry at about 4 and 12 kbar. These transitions conform with the criteria of polyhedral tilt transitions. (From Hazen and Finger, 1979, reproduced by permission of Gordon and Breach Science Publishers)

Talcs, micas and chlorites are common rock-forming minerals, but not all plausible compositions are observed in nature. A study of geometrical limits in the talc-like layer may be used to rationalize these unobserved compositions. The geometrical simplicity of these layer silicates has resulted in the prediction of structural details from unit-cell and cation radius data alone (Donnay $et\ al.$, 1964; McCauley and Newnham, 1971; Appelo, 1978). Predictions have been confirmed in the case of $KMg_3AlSi_3O_{10}(OH)_2$ (the mica phlogopite) at high temperature (Takeda and Morosin, 1975) and high pressure (Hazen and Finger, 1978a). The stability of these layer silicates is limited by the fit of adjacent octahedral and tetrahedral layers as measured by the tetrahedral rotation angle, α (Figure 10-5). It is observed that oxygen atoms within a layer are coplanar, that tetrahedra are regular, and that unit-cell $b = \sqrt{3}a$. Therefore,

$$\cos \alpha = 3\sqrt{3}d_o \sin \psi / 4\sqrt{2}d_t \qquad (10\text{-}1)$$

where d_o is the mean octahedral M–O distance, d_t is the mean tetrahedral T–O distance, and ψ is the octahedral 'flattening angle' which is approximately $59 \pm 1°$ in trioctahedral micas (Donnay $et\ al.$, 1964). It is not possible for α to be negative, and observations of Guidotti *et al.* (1975) suggest that rotation angles in excess of 12° are rare, and may be unstable for trioctahedral micas. From Hazen and Wones (1972, Figure 7):

$$\sin \psi \approx 1.154 - 0.144 d_o \qquad (10\text{-}2)$$

so the limits of α $(0° \leqslant \alpha \leqslant 12°)$ imply that the ratio of octahedral to tetrahedral M–O distances, d_o/d_t, is also limited to $1.235 \leqslant d_o/d_t \leqslant 1.275$. The effects of T, P and X on d_o/d_t are known, so it is possible to predict the value of α and consequently T–P–X regions of talc-like layer geometrical instability.

Tetrahedra of Si and Al are relatively rigid, and d_t is thus approximately invariant with T or P. Each tetrahedral layer composition, therefore, is characterized by critical maximum and minimum values of d_o, associated with the limiting values of $\alpha = 0°$ and 12°, respectively. The critical values of d_o for a

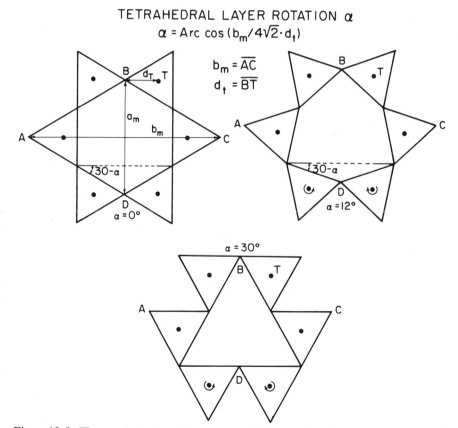

Figure 10-5. The tetrahedral rotation angle, α, in the talc-like layer of layer silicates. Tetrahedral rotations are a consequence of the different sizes of octahedral and tetrahedral layers in these silicates. (From Hazen and Wones, 1972, reproduced by permission of the Mineralogical Society of America)

given d_t may be attained through changes in temperature or pressure or octahedral layer composition, because all these variables alter d_o without significantly changing d_t.

Calculation of the upper temperature stability limits of trioctahedral micas ($\alpha = 0°$) has been presented by Hazen (1977b). In summary, micas of the composition $KR_3^{2+}AlSi_3O_{10}(OH)_2$ have a mean tetrahedral cation–anion distance of 1.649 A (Hazen and Burnham, 1973). The possible values of d_o for this d_t are, from equations (10-1) and (10-2), $2.035 \leqslant d_o \leqslant 2.110$ A. At room temperature and pressure, d_o for pure phlogopite is ≈ 2.06 A (Hazen and Burnham, 1973). For $^{VI}Mg^{2+}$–O bonds, mean thermal expansion is $\bar{\alpha} = 1.4 \times 10^{-5}C^{-1}$ and mean compression is $\bar{\beta} = 1.7 \times 10^{-4}$ kbar^{-1} (Chapters Six and Seven). Critical equations for the upper and lower geometrical stability of pure phlogopite are (T in °C, P in kbar)

$$2.060(1 - 0.00017P + 0.0000147T) \approx 2.110, \text{ or}$$
$$T \approx 12P + 1700 \text{ (upper limit)}, \tag{10-3}$$

and

$$2.060(1 - 0.00017P + 0.0000147T) \approx 2.035, \text{ or}$$
$$T \approx 12P - 900 \text{ (lower limit)}. \tag{10-4}$$

These $T–P$ lines represent upper and lower geometrical limits to phlogopite stability, based on the geometrical assumptions outlined above. They *do not* represent the actual $T–P$ stability of phlogopite, which melts significantly below temperatures of equations (10-3) and (10-4) (Yoder and Kushiro, 1969).

Critical $T–P$ lines for other end-member micas of the form $KR_3^{2+}AlSi_3O_{10}(OH)_2$ may be derived in a similar way with the radii of Shannon (1976): d_o for Ni, Co, Fe and Mn are 2.030, 2.085, 2.120 and 2.170 A, respectively. Values of $\bar{\alpha}$ and $\bar{\beta}$ from Chapters Six and Seven are used to obtain maximum and minimum equations for critical limits of end-member micas (T in °C, P in kbar):

	upper limit ($\alpha = 0°$)	lower limit ($\alpha = 12°$)
Ni	$T = 11P + 2800$	$T = 11P + 200$
Co	$T = 12P + 900$	$T = 12P - 1700$
Fe	$T = 13P - 300$	$T = 13P - 2900$
Mn	$T = 14P - 2000$	$T = 14P - 4600$

Divalent iron and manganese end-member trioctahedral micas are not stable under room conditions because the octahedral layer is too large. Nickel mica, on the other hand, has critically small octahedra and is predicted to have a tetrahedral rotation angle slightly greater than 12° under room conditions.

Ideal trioctahedral micas, $KR_3^{2+}AlSi_3O_{10}(OH)_2$, with large divalent octahedral cations (Fe^{2+}, Mn^{2+}) are geometrically unstable because d_o/d_t exceeds 1.275. Iron and manganese micas can be stabilized, however, by at least four common substitutional modifications:

(1) $^{IV}Al^{3+} + {}^{VI}Al^{3+} \rightleftharpoons {}^{IV}Si^{4+} + {}^{VI}R^{2+}$ (aluminum substitution)
(2) $^{VI}R^{2+} + (OH)^- \rightleftharpoons {}^{VI}R^{3+} + O^{2-}$ (oxidation)
(3) $^{IV}Al^{3+} \rightleftharpoons {}^{IV}Fe^{3+}$ (tetrahedral ferric iron substitution)
(4) $2\,^{VI}R^2 \rightleftharpoons Ti^{4+} + \square$ (titanium + vacancy substitution).

Each of these modifications causes a reduction in d_o/d_t and, consequently, an increase in fit between octahedral and tetrahedral layers of iron and manganese micas.

Consider mechanism (1) in detail. Micas of the form $K(R_{2.50}{}^{2+}Al_{0.50})$ $(Al_{1.50}Si_{2.50})O_{10}(OH)_2$, for which $d_t \approx 1.67$ A, have critical limits of $2.06 \leqslant d_o \leqslant 2.13$ A. The addition of 1/6 Al to the octahedral layer decreases mean octahedral M–O distances to approximately 2.00, 2.03, 2.05, 2.08 and 2.13 for divalent Ni, Mg, Co, Fe and Mn, respectively. If these values are used, critical upper and lower stability limits for several micas of this type are (T in °C, P in kbar):

	upper ($\alpha = 0°$)	lower ($\alpha = 12°$)
Ni	$T = 11P + 3400$	$T = 11P + 2100$
Mg	$T = 12P + 3400$	$T = 12P + 1000$
Co	$T = 12P + 2800$	$T = 12P + 300$
Fe	$T = 13P + 1700$	$T = 13P - 700$
Mn	$T = 14P$	$T = 14P - 2400$

For a given octahedral R^{2+} cation, aluminous micas with $(Al_{1.5}Si_{2.5})$ tetrahedral layer composition have critical geometrical limits calculated to be approximately 2000°C higher than for $(AlSi_3)$ layers. Biotites rich in Fe^{2+} and Mn^{2+} will thus be stabilized by this aluminum substitution. Nickel mica, on the other hand, will not be stabilized with excess aluminum because α will increase above 12°. A mica of composition $K(Ni_{2.5}Al_{0.5})(Al_{1.5}Si_{2.5})O_{10}(OH)_2$ is predicted to have an unstable tetrahedral rotation of $\approx 18°$ under room conditions.

Effects of octahedral cation oxidation are similarly dramatic. Pure annite, $KFe_3^{2+}AlSi_3O_{10}(OH)_2$, is predicted to be unstable; this ideal end member is not known in nature and has not been synthesized (Hewitt and Wones, 1975). 'Annites' of the composition $K(Fe_{2.7}{}^{2+}Fe_{0.3}{}^{3+})(AlSi_3)O_{10.3}(OH)_{1.7}$ have been synthesized, however. The values of d_o and d_t for this slightly oxidized mica are ≈ 2.105 and 1.65 A, giving $d_o/d_t = 1.275$, which is just within the possible range. This 'critical structure' mica is predicted to have $\alpha = 0°$.

Thus, although the actual stability limits of minerals with the talc-like layer are commonly less than those suggested by geometrical limits alone, the geometrical approach is successful in rationalizing some compositional limits of these layer silicates.

(2) The serpentine layer

Principles developed in the discussion of the talc-like layer may also be applied to the mineral serpentine, $(Mg_3Si_2O_5(OH)_4)$, in which a single tetrahedral layer interfaces with an octahedral layer (Figure 10-6). Like the talc example, the serpentine layer has limits of temperature, pressure and composition, beyond which a planar silicate layer cannot be formed. The serpentine layer, however, is able to accommodate conditions outside these limits by the remarkable mechanism of bending layers, much as a dimetallic thermostat coil responds to changes in temperature. Thus, the serpentine structure may have planar (lizardite), undulating (antigorite), and coiled or cylindrical (chrysotile asbestos) 'layers' which form as the result of varying degrees of octahedral and tetrahedral misfit (Figure 10-7).

For a given temperature, pressure and composition of formation, the serpentine layer has an equilibrium radius of curvature, R. Given a specific composition, any change in temperature or pressure will tend to alter the equilibrium curvature, because the tetrahedral layer has significantly smaller thermal expansion or compression than the octahedral layer. The tetrahedral layer has near zero $\bar{\alpha}$ and $\bar{\beta}$, whereas for the octahedral Mg layer $\bar{\alpha} = 1.4 \times 10^{-5}\,°C^{-1}$ and $\bar{\beta} = 1.7 \times 10^{-4}\,kbar^{-1}$. Assuming that the radius of curvature is R_0 at some T_0 and P_0, then the radius of curvature at T and P is given by:

$$R = \frac{D(1 + \bar{\alpha}\Delta T - \bar{\beta}\Delta P)}{(\bar{\alpha}\Delta T - \bar{\beta}\Delta P + D/R_0)} \tag{10-5}$$

as illustrated in Figure 10-8, where D is the distance between the cations in the octahedral and tetrahedral layers (approximately 3 A). If a serpentine that formed at 800°C and 3 kbar is quenched to room conditions, then the net curvature of an

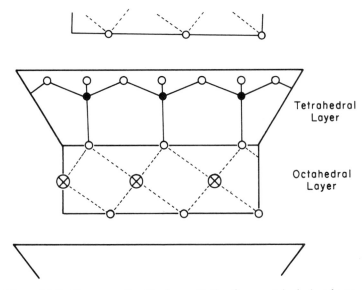

Figure 10-6. The serpentine-like layer. Units of one octahedral and one tetrahedral layer are bonded by Van der Waals forces

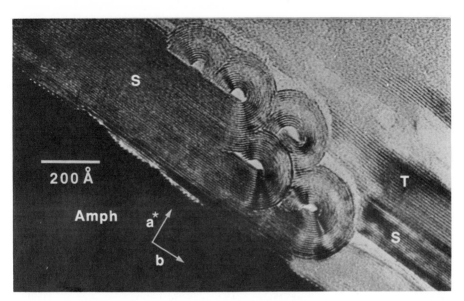

Figure 10-7. High-resolution transmission electron micrograph of serpentine. Note the variation in the curvature of the serpentine-like layers. (From Veblen, 1979, *Science* **206**, 1398–1400. Copyright 1979 by the American Association for the Advancement of Science)

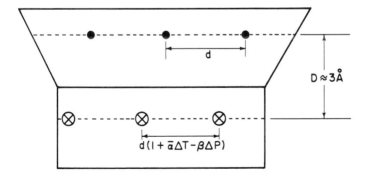

Figure 10-8. The radius of curvature of the serpentine-like layer is a function of temperature and pressure. The tetrahedral layer is approximately constant in size with changes in T and P, whereas the octahedral layer may undergo significant size variation. R is the radius of curvature of the layers after a change of conditions ΔT and ΔP, R_0 is the initial radius of curvature, D is the distance between the plane of tetrahedral and octahedral cations (approximately 3 A), and $\bar{\alpha}$ and $\bar{\beta}$ are the mean linear thermal expansion and compressibility of the octahedral layer, respectively

initially flat layer could be smaller than 300 A. In many chrysotiles, however, radii of 100 A and less are observed. It appears, therefore, that the initial composition of serpentine layers has the dominant control on the observed form. Furthermore, it seems likely that in mixed chrysotile–lizardite intergrowths (Figure 10-7) there may be compositional variations between these two forms of serpentine.

D. Martensitic transitions—chain silicates

Chain metasilicates in the system $(Mg, Fe, Ca, Mn)SiO_3$ exist in several structural modifications, as illustrated in Figure 10-9. All of these compounds have chains of corner-linked silicon tetrahedra connected by 6-, 7- and 8-coordinated R^{2+} polyhedra. The major differences between the several pyroxene and pyroxenoid structure types relate to the size and number of R^{2+} sites. The stable structure at any given $T–P–X$ appears to be governed in large measure by the relative size of silicon tetrahedra, which vary little with temperature or pressure, and the R^{2+} sites, which vary significantly with T, P and X. Pyroxenes have the smallest average R^{2+} polyhedral volume, followed by 7-, 5- and 3-repeat chain pyroxenoids (Abrecht and Peters, 1975). Phase boundaries between these different forms may approximate surfaces of constant R^{2+} volume in $T–P–X$ space (Simons, 1978).

Transformations from pyroxenes to pyroxenoids, as well as between different pyroxenoids, involve changes in topology, but extensive portions of the structure may remain intact. Pyroxene–pyroxenoid transitions may thus be considered as Martensitic transformations. In high-temperature experiments on $MnSiO_3$ (johannsenite) Morimoto et al. (1966) observed an oriented (i.e. single-crystal) transformation from pyroxene to the 3-repeat pyroxenoid (bustamite) at 900°C.

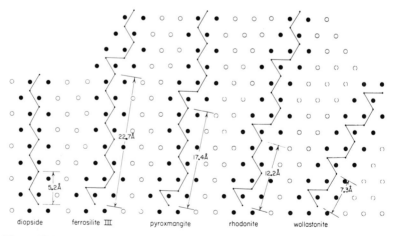

Figure 10-9. Pyroxene and pyroxenoid chain silicates. (From Burnham, 1966, *Science* **154**, 513–516. Copyright 1966 by the American Association for the Advancement of Science)

The increase of Mn^{2+} volume because of heating, and the corresponding decrease in ratio of tetrahedral to R^{2+} polyhedral size, is the driving force behind this phase transition. Similarly, pyroxenoids at high temperature and room pressure might convert to pyroxene at high pressure because of an increase in tetrahedral to R^{2+} cation size ratio.

III. POLYHEDRAL STABILITY FIELDS

Many geometrical limits to structure are the result of the *relative* sizes of structural elements, as demonstrated in the previous section. It is plausible that there also exist *absolute* limits to the sizes of certain components of a crystal. The specific components to be considered are cation–anion bond lengths, and cation coordination polyhedra of a fixed composition; therefore, temperature and pressure are the only intensive variables treated. It should be emphasized that no absolute size limits to bonds or polyhedra have been conclusively demonstrated, and what follows, consequently, is largely speculative.

A. Critical limits to bond distances

Are there absolute limits to bond distances in ionic solids? The concept of maximum and minimum distances for bonds within each type of cation polyhedron is reasonable. For example, octahedral Mg–O bonds could not exceed 2.5 A without dissociating; they could not compress to 1.5 A without a change in coordination number. At high temperature cation–anion bonds dissociate to some extent during melting. Even if the local cation environment is unchanged in melt formation, the cation–anion bonds will be disrupted at vaporization temperatures. Cation–anion bonds would appear to be disrupted owing to thermal vibrations at distances exceeding a few per cent of their room-temperature values.

There may also exist minimum stable bond distances for some cation polyhedra. It is well known that in high-pressure reconstructive transformations cations commonly increase in coordination, with a corresponding increase in cation–anion distance, and increased packing efficiency of atoms. For example, in all silicate minerals formed under crustal conditions, silicon is coordinated to four oxygens (^{IV}Si), but high-pressure transformations to phases with octahedral silicon (^{VI}Si) are well known (Table 10-2). If all silicon tetrahedra are assumed to have bulk moduli of approximately 2.5 Mbar (see Table 7-2), then it is possible to extrapolate structural characteristics of silicates to their high-pressure configurations near the $^{IV}Si = {}^{VI}Si$ transformations.

Calculations of mean ^{IV}Si–O distances in silicates at high pressure indicate that most of the known $^{IV}Si \rightarrow {}^{VI}Si$ transitions occur when the mean Si–O bond compresses to approximately 1.59 A (Hazen and Finger, 1978b). Framework silicates (e.g. SiO_2 and feldspars, $KAlSi_3O_8$) with relatively short Si–O bonds at room conditions (1.61 A) transform to ^{VI}Si compounds at pressures near 100 kbar. Chain silicates with ambient Si–O bond distances of approximately 1.63 A transform to orthosilicates with longer ^{IV}Si–O bonds or to ^{VI}Si compounds at

Table 10-2. High-pressure transitions in silicates from ^{IV}Si to ^{VI}Si forms

Formula	^{IV}Si Phase	^{VI}Si Phase	Transition Pressure	Transition d_{Si-O}	Reference
SiO_2	Coesite	Stishovite	80	1.59	Yagi and Akimoto (1976)
Al_2SiO_5	Kyanite	Corundum + Stishovite	160	1.60	Liu (1974)
$MgSiO_3$	Clinoenstatite	Ilmenite-type	250	1.59	Ito (1977), Liu (1977a)
$FeSiO_3$	Clinoferrosilite	Wustite + Stishovite	250	1.59	Liu (1976)
$CaSiO_3$	Wollastonite	Perovskite-type	160	1.60	Liu and Ringwood (1975)
$ZnSiO_3$	Zinc pyroxene	Ilmenite-type	180	1.59	Liu (1977a)
Mg_2SiO_4	Mg silicate spinel	Periclase + Perovskite-type	270	1.59	Ito (1977)
Fe_2SiO_4	Fe silicate spinel	Wustite + Stishovite	250	1.59	Liu (1976)
Ni_2SiO_4	Ni silicate spinel	NiO + Stishovite	190	1.61	Liu (1975a)
Co_2SiO_4	Co silicate spinel	CoO + Stishovite	180	1.61	Liu (1975b)
$NaAlSi_2O_6$	Jadeite	Calcium–Ferrite-type + Stishovite	180	1.59	Liu (1977b)
$KAlSi_3O_8$	Orthoclase	Hollandite-type	100	1.59	Ringwood et al. (1967)
$Mg_3Al_2Si_3O_{12}$	Pyrope	Ilmenite-type	245	1.59	Liu (1977c)

about 150 kbar. Orthosilicates, including olivine and silicate spinel forms of $(Mg, Fe)_2SiO_4$, have long Si–O bonds under room conditions (1.655 A) and transform to ^{IV}Si compounds at about 250 to 300 kbar. For each of these silicate types the average ^{IV}Si–O bond distance projected to the transformation pressure is 1.59 A, which therefore may be a minimum value.

Additional evidence for a critical minimum ^{IV}Si–O distance is given by Hill and Gibbs (1978), who found an inverse correlation between Si–O distance and Si–O–Si angle. At angles near the limiting value of 180° a minimum Si–O distance of approximately 1.59 A is observed. For bond lengths shorter than this critical value the repulsive energy terms, as modelled in equations 6-2 and 6-3, may become significant and thereby limit density increases. Transitions from ^{IV}Si to ^{VI}Si increase the Si–O separation as well as the packing efficiency.

B. Limits to polyhedral stability

An important conclusion of high-temperature and high-pressure crystallographic studies is that a cation–anion polyhedron has certain properties that are, to a first approximation, independent of linkages to other polyhedra. Polyhedral size, shape, expansivity and compressibility all appear to be functions of the primary coordination sphere of atoms in the great majority of structures. It is not unreasonable that absolute limits of temperature and pressure *stability* should be added to this list of polyhedral properties. If, as proposed above, there are limits to cation–anion bond distances, then polyhedral stability fields must be a consequence.

Consider the example of silicon in tetrahedral coordination. Data on high-pressure phase transitions are consistent with a minimum size for tetrahedral Si–O bonds of 1.59 A. This distance is achieved at room temperature at pressures of about 300 kbar or less in all measured silicates; therefore, 300 kbar may be an upper pressure limit for the silicon tetrahedron. Similarly, silicon in octahedral coordination is known primarily from high-pressure phases, synthesized above 80 kbar. Eighty kbar may thus be taken as a lower pressure limit for octahedral silicon (though quenched *metastable* phases with octahedral silicon persist at room conditions). Temperature has little effect on Si–O bond distances in either tetrahedra or octahedra, and tetrahedral species exist in silicate melts to very high temperatures (>3000 °C). In addition, $(\partial P/\partial T)$ of Si IV $\rightleftharpoons$ VI transitions are generally very small. A 'phase diagram' for silicon coordinated to oxygen thus would be of the form shown in Figure 10-10.

Each cation at ambient conditions has a limited number of known coordination states. Furthermore, phase equilibria studies have demonstrated that coordination number may change in regular and predictable sequence during high-temperature or high-pressure transformations. It is reasonable to conclude that each type of cation coordination group (e.g. ^{IV}Si or $^{VI}Fe^{3+}$) has a range of temperature–pressure conditions under which it may be found, and other T–P conditions under which only higher or lower coordination numbers are observed. Superposition of

214

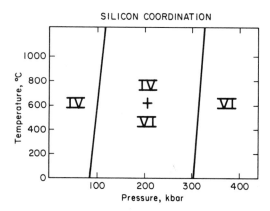

Figure 10-10. Polyhedral stability fields for $Si^{4+}-O^{2-}$ coordination groups

T–P polyhedral stability fields for several different cations may facilitate the prediction of the structural state and stability of crystals, glasses and liquids at conditions not yet attainable in laboratory experiments.

IV. CONCLUSIONS

One of the many contributing factors to phase transitions is geometrical instability that arises from the misfit of adjacent structural elements. A knowledge of continuous structural variations can lead to predictions of these discontinuities—phase transitions—providing limiting structural geometries are known. Geometrical limits may be described in terms of the relative sizes of two structural subunits, such as radii of two ions, volumes of two polyhedra, or size of a cavity filler compared to the cavity itself. It is also possible, though not proved, that absolute limits on bond distances or polyhedral stability may influence some phase transitions. The study of phase transitions and the geometrical limits to some structures promises to be one of the most rewarding fields for the application of high-temperature, high-pressure crystallography.

REFERENCES

Abrecht, J., and T. Peters (1975) Hydrothermal synthesis of pyroxenoids in the system $MnSiO_3-CaSiO_3$ at Pf = 2 kb. *Contrib. Mineral. Petrol.* **50**, 241–246.

Appelo, C. A. J. (1978) Crystal energy and layer deformation of micas and related minerals. I. Structural models for 1M and $2M_1$ polytypes. *Am. Mineral.* **63**, 782–791.

Boccara, N. (1968) Second-order phase transitions characterized by deformation of the unit cell. *Annals of Physics* **47**, 40–64.

Buerger, M. J. (1951) Crystallographic aspects of phase transformations. In *Phase Transformations in Solids*, R. Smoluchowski, J. E. Mayer, and W. A. Weyl (eds.), J. Wiley and Sons, NY, 183–209.

Buerger, M. J. (1972) Phase Transformations. *Soviet Phys. Crystallography* **16**, 959–968.

Burnham, C. W. (1966) Ferrosilite III: a triclinic pyroxenoid-type polymorph of ferrous metasilicate. *Science* **154**, 513–516.

Clarke, R., and L. Benguigui (1977) The tricritical point in $BaTiO_3$. *J. Phys. C-Solid State Phys.* **10**, 1963–1973.

Cohen, L. H., and W. Klement (1976) Effect of pressure on reversible solid-solid transitions in nepheline and carnegieite. *Mineral Mag.* **40**, 487–492.

Cohen, L. H., W. Clement and H. G. Adams (1974) Yet more observations on the high-low quartz inversion. *Am. Mineral.* **59**, 1099–1104.

Deer, W. A., R. A. Howie and J. Zussman (1963) *Rock-Forming Minerals*, John Wiley and Sons, NY, Vol. **IV**, 338–350.

Donnay, G., J. D. H. Donnay and H. Takeda (1964) Trioctahedral one-layer micas. II. Prediction of the structure from composition and cell dimensions. *Acta Crystallogr.* **17**, 1374–1381.

Guidotti, C. V., J. T. Cheney and P. D. Conatore (1975) Interrelationship between Mg/Fe ratio and octahedral Al content in biotite. *Am. Mineral.* **60**, 849–853.

Hazen, R. M. (1976) Sanidine: predicted and observed monoclinic-to-triclinic reversible transformations at high pressure. *Science* **194**, 105–107.

Hazen, R. M. (1977a) Mechanisms of transformation and twinning in gillespite at high pressure. *Am. Mineral.* **62**, 528–533.

Hazen, R. M. (1977b) Temperature, pressure, and composition: structurally analogous variables. *Phys. Chem. Minerals* **1**, 83–94.

Hazen, R. M., and C. W. Burnham (1973) The crystal structures of one-layer phlogopite and annite. *Am. Mineral.* **58**, 889–900.

Hazen, R. M., and L. W. Finger (1978a) The crystal structures and compressibilities of layer minerals at high pressure. II. Phlogopite and chlorite. *Am. Mineral.* **63**, 293–296.

Hazen, R. M., and L. W. Finger (1978b) Crystal chemistry of silicon-oxygen bonds at high pressure: implications for the Earth's mantle mineralogy. *Science* **201**, 1122–1123.

Hazen, R. M., and L. W. Finger (1979) Polyhedral tilting: a common type of pure displacive phase transition and its relationship to analcite at high pressure. *Phase Transitions* **1**, 1–22.

Hazen, R. M., and D. R. Wones (1972) The effect of cation substitution on the physical properties of trioctahedral micas. *Am. Mineral.* **57**, 103–129.

Hazen, R. M., and D. R. Wones (1978) Predicted and observed compositional limits of trioctahedral micas. *Am. Mineral.* **63**, 885–892.

Hewitt, D. A., and D. R. Wones (1975) Physical properties of some synthetic Fe–Mg–Al trioctahedral micas. *Am. Mineral.* **60**, 854–862.

Hill, R. J., and G. V. Gibbs (1979) Variations in $d(T–O)$, $d(T \ldots T)$ and $\langle TOT$ in silica and silicate minerals. *Acta Crystallogr.* **B35**, 25–30.

Ito, E. (1977) The absence of oxide mixture in high-pressure phase of Mg-silicate. *Geophys. Res. Lett.* **4**, 72–74.

Langer, K., and W. Schreyer (1969) Infrared and powder x-ray diffraction studies of cordierite. *Am. Mineral.* **54**, 1442–1459.

Liu, L.-G. (1974) Disproportionation of kyanite to corundum plus stishovite at high pressures and temperatures. *Earth Planet. Sci. Lett.* **24**, 224–228.

Liu, L.-G. (1975a) Disproportionation of Ni_2SiO_4 to stishovite plus bunsenite at high pressures and temperatures. *Earth Planet. Sci. Lett.* **24**, 357–362.

Liu, L.-G. (1975b) High-pressure disproportionation of Co_2SiO_4 spinel and implications for Mg_2SiO_4 spinel. *Earth Planet. Sci. Lett.* **25**, 286–290.

Liu, L.-G. (1976) High-pressure phases of $FeSiO_3$ with implications for Fe_2SiO_4 and FeO. *Earth Planet. Sci. lett.* **33**, 101–106.

Liu, L.-G. (1977a) Post-ilmenite phases of silicates and germanates. *Earth Planet. Sci. Lett.* **35**, 161–168.

216

Liu, L.-G. (1977b) High-pressure $NaAlSiO_4$: the first silicate calcium ferrite isotype. *Geophys. Res. Lett.* **4**, 183–186.

Liu, L.-G. (1977c) First occurrence of the garnet-ilmenite transition in silicates. *Science* **195**, 990–991.

Liu, L.-G., and A. E. Ringwood (1975) Synthesis of a perovskite-type polymorph of $CaSiO_3$. *Earth Planet. Sci. Lett.* **28**, 209–211.

McCauley, J. W., and R. E. Newnham (1971) Origin and prediction of ditrigonal distortions in micas. *Am. Mineral.* **56**, 1626–1638.

Megaw, H. D. (1973) *Crystal Structures: A Working Approach*, W. B. Saunders, Philadelphia.

Megaw, H. D. (1974) The seven phases of sodium niobate. *Ferroelectrics* **7**, 87–89.

Merrill, L., and W. A. Bassett (1975) The crystal structure of $CaCO_3$ (II), a high-pressure metastable phase of calcium carbonate. *Acta Crystallogr.* **B31**, 343–349.

Morimoto, N., K. Koto and T. Shinohara (1966) Oriented transformation of johannsenite to bustamite. *Mineralogical J. (Japan)* **5**, 44–64.

Pauling, L. (1939) *The Nature of the Chemical Bond and the Structure of Molecules and Crystals* Cornell Univ. Press, Ithaca, NY, 1st ed.

Ringwood, A. E., A. Reid and A. Wadsley (1967) High-pressure $KAlSi_3O_8$, an aluminosilicate with sixfold coordination. *Acta Cryst.* **23**, 1093–1095.

Rosenhauer, M., and H. K. Mao (1975) Studies on the high-pressure polymorphism of analcite by power x-ray diffraction and differential thermal analysis methods. *Carnegie Inst. Washington Year Book* **74**, 413–415.

Shannon, R. D. (1976) Revised effective ionic radii and systematic studies of interatomic distances in halides and chalcogenides. *Acta Crystallogr.* **A32**, 751–767.

Simons, B. (1978) *Pyroxen-Pyroxenoid-Unwandlungen im System $FeSiO_3$-$MnSiO_3$-$MnGeO_3$-$FeGeO_3$ unter hohen Drucken und hohen Temperaturen.* Ph.D. Thesis, Reinisch-Westfalische Technische Hochschule Aachen.

Smyth, J. R. (1974) Experimental study on the polymorphism of enstatite. *Am. Mineral.* **59**, 345–352.

Strens, R. G. J. (1967) Symmetry–entropy–volume relationships in polymorphism. *Mineral. Mag.* **36**, 565–577.

Takeda, H., and B. Morosin (1975) Comparison of observed and predicted structural parameters of mica at high temperature. *Acta Crystallogr.* **B31**, 2444–2452.

Veblen, D. (1979) Serpentine minerals: intergrowths and new combination structures. *Science* **206**, 1398–1400.

Yagi, T., and S. Akimoto (1976) Direct determination of coesite-stishovite transition by in-situ x-ray measurements. *Tectonophysics* **35**, 259–270.

Yoder, H. S., and I. Kushiro (1969) Melting of a hydrous phase: phlogopite. *Am. J. Sci.* **267A**, 558–582.

Author Index

Abrahams, S. C., 6
Abrecht, J., 210
Ackermann, R. J., 126
Adams, H. G., 198
Agrell, S., 198
Akimoto, S., 154, 212
Anderson, D. L., 151
Anderson, O. L., 151, 154, 158, 162
Andre, D., 59
Andrews, K. W., 125, 126, 127
Apello, C. A. J., 204
Austin, J. B., 125, 126

Bailey, S. W., 174
Barnett, J. D., 37, 38, 39, 58, 68, 190
Barns, R. L., 124
Bassett, W. A., 21, 22, 23, 26, 27, 28, 29,
    30, 31, 32, 33, 34, 35, 41, 51, 58, 59,
    60, 150, 154, 190, 197
Bassi, G., 50, 51
Batchelder, D. N., 69
Baur, P., 59
Baur, W., 192
Beals, R. J., 125, 126
Bell, P. M., 20, 25, 28, 29, 38, 39, 40
Belson, H. S., 127
Benguigui, L., 197
Bernstein, J. L., 6
Beyerlein, R. A., 151
Billy, M., 151
Birch, F., 149, 154
Bish, D. L., 192
Bishop, F. C., 172
Block, S., 21, 22, 37, 38, 39, 51, 58, 59, 68,
    151

Bobrovskii, A. B., 125
Boccara, N., 199
Braun, A. B., 21, 151
Bridgman, P. W., 151, 188
Brooksbank, D., 125, 126, 127
Brown, G. E., 6, 8, 15, 128, 129, 131, 132,
    133
Buchanan, J. A., 53
Buerger, M. J., 195
Bukowinski, M. S. T., 149
Bunting, E. N., 19
Burnham, C. W., 23, 80, 117, 119, 132,
    151, 174, 192, 206, 210
Busing, W. R., 46, 50, 82, 122

Cahn, J. W., 19
Cameron, M., 124, 129, 130, 132, 133,
    140, 141
Campbell, W. J., 125, 126
Carter, R. E., 125
Cartz, L., 151
Chang, T. S., 126
Cheney, J. T., 205
Chhabildas, L. C., 160
Chrenko, R. M., 20
Clarke, R., 197
Clement, W., 198
Clendennen, R. L., 154, 155
Cohen, L. H., 198
Conatore, P. D., 205
Cook, R. L., 125, 126
Cox, D. E., 132

Dale, H., 168
D'Amour, H., 51, 150

217

Das, C. D., 120, 149
Dayal, B., 125
Decker, D. L., 190
Deer, W. A., 202
Deganello, S., 124
Dempsey, M., 192
Denner, W., 51, 150
DeVries, R. C., 20, 58
Dollase, W. A., 192
Donnay, G., 204
Donnay, J. D. H., 204
Drickamer, H. G., 154, 155
Dunn, K. J., 20

Eisenstein, A. J., 125
Elliott, R. O., 126
Ettenberg, M., 127

Feder, R., 127
Finkel, V. A., 125
Foit, F. F., Jr., 5, 6, 131
Foreman, N., 131
Forman, R. A., 38
Fourme, R., 25, 26, 58, 59
Fraase Storm, G. M., 6

Gatos, H. C., 124
Ghose, S., 82, 83, 130, 131, 149, 151, 155
Gibbs, G. V., 86, 131, 133
Glazer, A. M., 6, 129
Goldschmidt, H. J., 5
Grain, C. F., 125, 126
Green, A. T., 125, 126
Guggenheim, H. J., 124
Guidotti, C. V., 205

Hadidiacos, C. G., 6, 9, 11, 60, 62
Hall, H. T., 190
Hamilton, W. C., 46, 87
Hanneman, R. E., 124
Hewitt, D. A., 207
Hill, R. J., 213
Hilliard, J. E., 19
Ho, P. S., 69
Hochella, M. F., 131, 133
Holzapfel, W. B., 27, 59, 150
Horn, M., 128
Houston, B., 127
Howie, R. A., 202
Hyde, B. G., 168

Ishizawa, N., 6, 15
Ito, E., 212

Iwai, S., 6, 15

Jamieson, J. C., 19, 20
Jayalakshmi, K., 6
Johnson, C. K., 86, 87, 88
Jorgensen, J. D., 151, 154
Jun, C. K., 127

Kardashev, B. K., 69
Karplus, M., 161
Kartmazov, G. N., 125
Kas'kovich, N. S., 69
Kasper, J. S., 19
Keer, H. V., 120, 149
Keller, R., 27, 59, 150
Kempter, C. P., 126
King, H. E., 39, 46, 51, 52, 150, 161
Kittel, C., 117, 119
Koster, A. S., 132
Koto, K., 210
Kumar, S., 120, 124, 127
Kushiro, I., 206
Kvaskov, V. B., 125, 126

Labbe, L. C., 151
Lager, G. A., 128, 129, 133, 141
Langer, K., 198
Larionov, A. L., 69, 127
Lawson, A. W., 18, 20
Leopold, M. H., 125
Levien, L., 150, 151, 155, 183, 185
Levy, H. A., 46, 82, 122
Light, T., 127
Lines, M. E., 124
Lippincott, E. R., 19
Lissalde, F., 6
Liu, L. G., 154, 212
Lonsdale, K., 124
Lovell, G. H. B., 125, 126
Lynch, R. W., 6, 154

Mabud, S. A., 129
Malkin, B. Z., 69, 127
Mao, H. K., 20, 25, 28, 29, 38, 39, 40, 202
McCauley, J. W., 204
Meagher, E. P., 128, 129, 133, 141
Megaw, H. D., 120, 195, 198, 199
Meier, W. M., 192
Merrill, L., 21, 22, 23, 26, 27, 28, 29, 30,
    31, 32, 33, 34, 35, 41, 51, 59, 60,
    150, 190, 197
Minato, I., 6, 15
Ming, L. C., 58, 59

Min'ko, N. I., 125, 126
Miyano, S., 58
Miyata, T., 6, 15
Moore, M. J., 58
Morimoto, N., 210
Morosin, B., 6, 131, 133, 204
Munro, D. C., 20

Nachtrieb, N. D., 20
Nafe, J. E., 151
Newnham, R. E., 204
Newton, R. C., 172
Nicholson, W. L., 53
Nielsen, T. H., 125
Nikanorov, S. P., 69

O'Bryan, H. M., 124
Ohashi, Y., 6, 8, 9, 11, 60, 62, 80, 92, 103,
    117, 119, 139
Okamura, F. P., 82, 83, 130, 131
O'Keeffe, M., 168

Paff, R. J., 127
Papike, J. J., 6, 7, 124, 130, 132, 133, 140,
    141
Pathak, P. D., 126, 127
Pauling, L., 1, 85, 115, 117, 137, 166, 196
Peacor, D. R., 5, 6, 128, 131, 132
Pease, R. S., 127
Peters, T., 210
Peterson, R. C., 85
Petukhov, V. A., 126
Phillips, V. A., 19
Piermarini, G. J., 21, 22, 37, 38, 39, 51, 58,
    59, 68, 151
Pillars, W. W., 132
Pinckney, L., 185
Porter, R. N., 161
Prewitt, C. T., 6, 7, 8, 14, 15, 39, 124, 128,
    129, 130, 132, 133, 136, 140, 141,
    150, 151, 154, 155, 156, 166, 168,
    181, 182, 183, 184, 185
Prince, E., 53, 87, 88
Pryor, A. W., 87

Ralph, R. L., 51, 149, 151, 155
Rao, K. V. K., 186
Rao, R. V. G., 120, 149
Reeber, R. R., 6
Reid, A. F., 212
Renaud, J. P. P., 132
Renaud, M., 59
Ribbe, P. H., 86

Rice, C. E., 128
Rieck, G. D., 132
Riedner, R. J., 151
Rigby, G. R., 125, 126
Ringwood, A. E., 212
Robinson, K., 86
Robinson, W. R., 128
Rooymans, C. J. M., 20
Rosenhauer, M., 202
Ross, F. K., 131, 133
Ross, M., 6, 7
Ruoff, A. L., 69, 160

Santoro, A., 51
Sato, Y., 154
Schiferl, D., 23, 24, 27, 34, 59, 150
Schreiber, E., 154
Schreyer, W., 198
Schuele, D. E., 69
Schulz, H., 51, 150
Schwerdtfeger, C. F., 128
Shannon, R. D., 166, 167
Sharma, S. S., 69, 126
Shinohara, T., 210
Shoemaker, D. P., 50, 51
Shomaker, V., 87
Simmons, G., 149, 154
Simmons, R. O., 69
Simons, B., 210
Singh, H. P., 125
Skinner, B. J., 125, 126, 127
Slater, J. C., 166
Sleight, A. W., 132
Smith, J. V., 174
Smyth, J. R., 6, 7, 12, 128, 129, 132, 198
Sorenson, D. B., 58
Sorrell, C. A., 126
Srinivasa, S. R., 151
Stecura, S., 126
Stewart, D. B., 101, 102
Strakna, R. E., 127
Strens, R. G. J., 192, 199
Sueno, S., 6, 8, 15, 58, 124, 129, 130, 132,
    133, 140, 141
Sung, C. M., 58
Suzuki, I., 125
Swanson, D. K., 85

Takahashi, T., 58, 154
Takeda, H., 131, 133, 204
Tang, T. Y., 18
Taylor, M., 129
Thompson, J. B., 172

220

Tichý, K., 50
Tossell, J. A., 149
Trivedi, J. M., 126, 127
Trucano, P., 126
Trueblood, K. N., 87
Truinstra, F., 6

Valeev, K. S., 125, 126
Van Uitert, L. G., 124
Van Valkenburg, A., 19, 21, 58, 59
Vasavada, N. G., 126, 127
Veblen, D., 209
Vegard, L., 168
Villiger, H., 192
Viswamitra, M. A., 6
Von Limbach, D., 101, 102

Wadsley, A. D., 212
Wang, H., 149, 154
Wechsler, B. A., 129

Weidner, D. J., 150, 183
Weir, C. E., 19, 21, 22, 51, 59, 151
White, G. K., 120, 121
Willis, B. T. M., 87
Winter, J. K., 82, 83, 130, 131
Wones, D. R., 170, 171, 203, 205, 207
Wong, C., 69
Wood, B. J., 172
Worlton, T. G., 151
Wyckoff, R. W. G., 5

Yagi, T., 129, 150, 154, 183, 188, 189,
    190, 212
Yoder, H. S., Jr., 206
Young, R. A., 183

Zollweg, R. J., 125
Zou, G., 29, 38, 40
Zussman, J., 202
Zydzik, G., 124

Subject Index

Absorption corrections
  boron carbide, 74
  importance of, 14
  pressure cells, 51
Albite
  inverse relationship, 184
Alcohols
  pressure fluid, 37, 67
Alignment
  diamonds, 26
Alkali halides
  inverse relationship, 182
Analcite
  phase transformation, 201–202, 204
  polyhedral tilting, 201–202, 204
Anharmonic vibrations
  thermal parameters, 87, 88
Argon
  pressure calibration, 40
  pressure fluid, 38
Automation
  diffractometer, 10
  high-temperature control, 9–10
Averaging
  intensity data, 52

Beryllium
  diamond supports, 30–32
  pre-stressed, 31
  use in lever-arm cell, 21, 23
  X-ray diffraction effects, 43, 44, 52
Bond angle variance
  Definition, 86
Bond distances
  compositional effects, 173–174

compressibility, 149
critical limits, 211, 213
errors, 84–85
mean, 84
silicon–oxygen, 82, 83, 211–213
thermal corrections, 82–83
variations in, 82
Bonding
  aluminum–oxygen, 174
  ionic, 115
  iron–oxygen, 173
  magnesium–oxygen, 155, 173
  pressure effect on, 148
  silicon–oxygen, 155, 174
Boron carbide
  absorption, 74
  use in heated pressure cell, 60, 64
Bridgman, Percy
  high-pressure research, 151
Bulk modulus
  cesium chlorate, 158
  definition, 148
  halides, 156
  molar volume relationship, 151
  oxides, 156
  polyhedra, 152–156, 158, 180
  polyhedral linkages, 156
  polyhedral volume relationship,
    152–154, 156–158
  pressure derivatives, 159
  silicates, 156

Calcite, *see* calcium carbonate
Calcium carbonate
  calibration, 70–72

Calcium carbonate (*contd.*)
  double internal standards, 186
  equation of state, 70–72
  high-pressure crystal structure, 150
  high-pressure research, 19, 23
Calcium fluoride
  calibration, 69, 70
  equation of state, 69, 70
  X-ray diffraction at P and T, 73
Calibration
  calcium carbonate, 70–72
  calcium fluoride, 69–70
  diamond cells, 68–72
  difficulties, 68
  double internal standard, 70–72, 186
  fixed points, 14, 15
  high pressure, 38, 39
  high temperature, 6, 9, 10, 14
  internal standard, 15, 39, 68–70
  laser, use in, 39
  pressure cells, 68–72
  pressure at high temperature, 68–72
  ruby, 38
  temperature at high pressure, 68–72
Capillary mount
  high temperature, 13, 14
Cesium chloride
  bulk modulus, 158
Chicago, University of
  high-pressure research, 18, 19
Chimney effect
  corrections for, 10
Circuit diagram
  microheater control, 61, 62
  programmable temperature controller,
    11
Clinopyroxene
  high-pressure crystal structure, 151
  high-temperature research, 6
Coesite
  high-pressure crystal structure, 150
Composition
  coefficient of expansion defined, 165
  effect on bond distance, 173
  effect on compressibility, 190–191
  structural analogy with T and P, 178
Compositional expansion
  definition, 165
  effect of pressure, 191
Compressibility
  bond distances, 149
  calculation from bonding models, 149
  calculation from site potential, 148

  definition, 147
  effect of composition on, 190–191
  effect of pressure on, 189–190
  effect of temperature on, 187–189
  fixed structures, 149
  ionic bonds, 148
  linear, 147
  polyhedra, 85, 149, 181
  polyhedral linkages related to, 138
  volume, 147
Computer program
  poyhedral volumes and distortions, 103
  strain ellipsoid, 92
Cone axis photography
  gillespite, 45
Coordination number
  silicon, 211–213
Corundum
  high-pressure crystal structure, 150
Crystal centring
  diffractometry procedures, 46–49
Crystal mount
  compound, 36–37
  high-temperature silica capillary, 12–13
Crystal structure
  compressibility, 158
  high pressure, 149–151
  mica, 169
  prediction, 192
  pyroxene, 210
  pyroxenoids, 210
  serpentine, 207, 208
Crystallography
  comparative, 1
  high pressure, 17
  high temperature, 5

Diamond
  advantages for pressure research, 19
  alignment, 26, 30, 34, 35, 65
  Anvil design, 31–32
  breakage, 23, 27, 30, 34
  graphitization, 58
  mounting, 34, 64
  opposed-anvil configuration, 19–21, 35
  oxidation, 58
  pressure cell, 18, 19
  X-ray diffraction effects, 43, 44, 52
Diamond cell
  assembly of high temperature, 65–66
  calibration, 68–72
  calibration of temperature, 68–72
  crystal mounting, 66

design for heated, 60, 61
heated, 74
heaters, 57–66, 73
high-temperature X-ray photography, 72
history, 57–58
operation at high temperature, 64, 67
Diffractometer
four circle, 10
heated pressure cell, 67, 74
Diffractometry
crystal centring, 46–49
geophysical laboratory, 46
heated pressure cell, 74–75
high-pressure research, 46
intensity collection, 51
systematic errors, 46
Diopside
high-pressure crystal structure, 151
Distance least squares
procedures, 192
Distortions
angle variance, 86
calculation, 103, 108
polyhedra, 86, 103, 160
quadratic elongation, 86

Enstatite
high-pressure crystal structure, 149, 151
Equation of state
Birch, 160
Birch–Murnaghan, 160
calcium carbonate, 70–72
calcium fluoride, 69–70
empirical, 160
polynomial, 160
structural, 1
Errors
bond distances, 84, 85
propagation of, 78–79
systematic in diffractometry, 46

Fayalite
high-pressure crystal structure, 150
Feldspar
inverse relationship, 182, 184
isostructural surfaces, 180
phase transformation, 202–203
polyhedral tilting, 202–203
Fixed points
calibration, 14, 15

Fixed structures
compressibility, 149
Flame heaters
limitations, 7
oxy-hydrogen, 6
Fluorite, *see* Calcium fluoride
Forsterite
high-pressure crystal structure, 149, 150
Future prospects
high-pressure research, 53
high-temperature research, 15

Garnet
high-pressure crystal structure, 151
Gas-flow heaters
design, 7
Gaskets
centring, 36
cycling, 38
deformation, 23, 33, 38
design, 33, 36
development of the technique, 21
failure, 34
heated pressure cell, 66
heaters, 58
high temperature, 60
high-temperature behaviour, 58
inconel, 33
mounting, 36
steel, 33
Geometrical structure limits
layer silicates, 205
Geophysical Laboratory
diffractometry, 46
heated pressure cell, 74
high-pressure research, 40, 43
high-temperature research, 5
programmable temperature controller, 11
radiative heaters, 8
Gillespite
cone axis photography, 45
high-pressure crystal structure, 151
high-pressure research, 23
precession photography, 44
Glycerin
pressure fluid, 37
Goniometer head
heated pressure cell, 73
high pressure, 40–41
high temperature, 13
reference coordinate system, 47

Halides
  bulk moduli, 156
Harmonic vibrations
  thermal parameters, 86, 87
Heaters
  diamond cell, 57–60, 63
  gas flow, 7
  gaskets, 58
  miniature, 60
  open flame, 6
  pressure cells, 57–60
  programmable controller, 10–11
  radiative, 8–9
  single crystal, 5–6
  types, 6
  Weissenberg geometry, 5
  X-ray cameras, 5
Helium
  pressure fluid, 25
High pressure
  calibration, 38, 39
  crystallography, 17
  goniometer head, 40
  powder diffraction, 18, 20
High-pressure research
  Bridgman, Percy, 151
  Chicago, University of, 18, 19
  crystal structures, 150–151
  diffractometry, 46
  future prospects, 53
  geophysical laboratory, 40, 43
  gillespite, 23
  laser, 39
  National Bureau of Standards, 19, 20,
      21, 23, 38, 51
  precession camera, 21, 25, 42, 44
  safety, 40
High-temperature research
  future prospects, 15
  Geophysical Laboratory, 5
  olivine, 12
  pyroxene, 6
  thermocouple, 14
History
  diamond cells, 57–58
Hydrogen
  pressure fluid, 25, 38

Inconel
  gaskets, 33
Insulation
  mica, 65
  pyrophyllite, 63, 64

Internal standard
  double, 186
Inverse relationship
  alkali halides, 182
  counter examples, 185
  definition, 180
  examples, 183
  feldspar, 182, 184
  ideal case, 182
  pyroxenoid, 182, 185
  quartz, 183, 184
  rutile, 186
  simple structures, 181
  spinel, 182, 183
Ionic bonds
  compressibility, 148
  definition, 115
Ionic radii
  effects on molar volume, 168
  oxygen, 166
  ratio limits, 196
  relative, 168
  Shannon and Prewitt, 166–167
Ionicity
  empirical term, 157
Isochoric surfaces
  $T–P–X$ space, 179
Isostructural surfaces
  characteristics, 180
  feldspar, 180
  oxides, 179
  phase transformation, 200
  slopes, 180
  spinel, 179

Kapton
  thermal shielding, 9

Laser
  diamond cell heating, 59
  high-pressure research, 39
  single-crystal heating, 15
Layer silicates
  geometrical structure limits,
      205
  phase transformation, 203, 204, 205,
      207
Least-squares refinement
  data averaging, 52
  distance, 192
  robust resistant, 53
  thermal parameters, 88
  unit-cell dimensions, 50

Magnesium
  bond compressibility, 155
Methane
  high-pressure research, 38
Mica
  crystal structure, 169
  high-pressure crystal structure, 149, 151
  phase transformation, 205
  stability, 205, 207
  use in heated pressure cell, 65
  Vegard's law, 168, 170, 171
Mixing
  volume of, 171–172

National Bureau of Standards
  diamond cell heating, 58, 59
  high-pressure research, 19, 20, 21, 23, 38, 51
Neon
  pressure calibration, 40
  pressure fluid, 29, 38
Nitrogen
  use in heaters, 7

Ohashi, Y.
  program to calculate strain tensor, 92
  radiative heater, 8–9
Olivine
  high-pressure crystal structure, 150
  high-temperature research, 12
Orientation photography
  precession camera, 43
Oxides
  bulk modulus, 156
  isostructural surfaces, 179
  Vegard's law, 168, 169
Oxygen
  ionic radii, 166

Phase transformation
  analcite, 201–202, 204
  classification, 194–195, 197
  coordination change, 214
  displacive, 195
  feldspar, 202–203
  geometrical limits, 196
  isostructural surfaces, 200
  layer silicates, 203–205, 207
  martensitic, 195, 210
  mechanisms, 196
  mica, 205
  polyhedral tilting, 197–199, 201

pyroxene, 210–211
pyroxenoids, 210–211
radius ratio, 196
reconstructive, 195, 203
reversible, 195
silicon coordination change, 211–213
symmetry, 198
topology, 194–195, 197
twinning, 200–201
Phlogopite
  high-pressure crystal structure, 151
Platinum
  thermocouple, 12, 15
Pneumatic cell
  pressure cells, 25
Polyhedra
  anomalous compressibility, 158
  bulk modulus, 152–156, 158, 160
  compositional effects on volume, 173–175
  compressibility, 85, 149
  compressibility related to structure, 158
  distortions, 86, 103, 108, 160
  excess volume of mixing, 175
  fundamental unit of structure, 178
  linkages related to compression, 138, 156
  properties, 155, 178
  ratio of expansion to compression, 181
  size ratios, 178
  stability fields, 211, 213–214
  thermal expansion, 85, 180
  volume calculation, 103, 108
  volumes, 85
Polyhedral tilting
  analcite, 201–202, 204
  characteristics, 197
  definition, 197
  examples, 198, 201
  feldspar, 202–203
  phase transformation, 197–199, 201
  symmetry, 199
  twinning, 200–201
Polyhedral volume
  changes with T, P and X, 178
  ratios, 178
Powder diffraction
  high-pressure, 18, 20
Precession camera
  cone-axis photography, 43, 45
  heated pressure cell, 72–73
  high-pressure research, 21, 25, 42, 44
  orientation photography, 43

Pressure
    effect on bonding, 148
    effect on compositional expansion, 191
    effect on compressibility, 189
    effect on crystal structure, 138
    effect on thermal expansion, 187–189
    effect on thermal factors, 161
    inverse relationship with temperature, 180
    structural analogy with T and X, 178
Pressure calibration
    argon, 40
    internal, 39–40
    need for improvement, 53
    neon, 40
    Ramon spectroscopy, 39
    ruby, 39
    sodium chloride, 40
Pressure cells
    absorption corrections, 51
    calibration, 68–72
    cryogenically loaded, 28–29
    design, 30
    diamond, 18–19
    diamond cell design, 23
    Fourme, 25–26
    heated, 74
    heaters, 57–60
    high temperature, 65
    lever arm, 19, 21, 22, 25
    Megabar, 25
    Merrill and Bassett, 22, 26–27, 29, 30
    operation, 33
    pneumatic, 25–26
    sample mounting, 33, 36–37
    Schiferl, 24, 27–28
    split diamond, 18
    Stuttgart, 27
    temperature calibration, 68–72
    transverse geometry, 24, 27–28
    x-ray absorption, 51
Pressure fluids
    alcohols, 37, 67
    argon, 38
    gases, 38
    glycerin, 37
    helium, 25
    hydrogen, 25, 38
    hydrostatic, 29
    leakage, 36
    neon, 29, 38
    use in heated cell, 67
Pyrophyllite
    use in heated pressure cell, 63, 64

Pyroxene
    crystal structure, 210
    high-pressure crystal structure, 151
    high-temperature research, 6
    phase transformation, 210–211
Pyroxenoids
    crystal structure, 210
    inverse relationship, 182, 185
    phase transformation, 210–211

Quadratic elongation
    definition, 86
Quartz
    high-pressure crystal structure, 150
    inverse relationship, 183–184

Radiative heaters
    design, 8–9
    geophysical laboratory, 8
    Ohashi, Y., 8–9
    Stony Brook (SUNY), 8
    thermocouple, 9
Radii
    bonded, 168
    ionic, 166, 168, 196
    nonbonded, 168
Radius ratio
    phase transformation, 196
Raman spectroscopy
    pressure calibration, 39
Resistance heaters
    platinum element, 7, 8
Rigid-body motion
    thermal parameters, 87
Robust-resistant
    least-squares refinement, 53
Ruby
    fluorescence pressure calibration, 38–39
    high-pressure crystal structure, 150
Rutile
    high-pressure crystal structure, 150
    inverse relationship, 186

Safety
    high-pressure research, 40
Serpentine
    crystal structure, 207, 208
    layer bending, 208
    radius of curvature, 208–210
    transmission electron microscopy, 209
    varieties, 208
Silica glass
    high-temperature crystal mounting, 12–13

Silicates
    bulk modulus, 156
    chain, 210
    layer, 203–210
Silicon
    bond compressibility, 155
    bonding to oxygen, 211
    coordination number, 211–213
    polyhedral stability fields, 214
Sodium chloride
    high-pressure crystal structure, 150, 161
    high-temperature calibration, 15
    pressure calibration, 40
Solid solution
    iron–magnesium, 173
    silicon–aluminum, 174
Spinel
    high-pressure crystal structure, 150
    inverse relationship, 182, 183
    isostructural surfaces, 179
Stability
    geometrical limits, 196
    polyhedra, 211–214
Stability fields
    polyhedra, 211–214
Steel
    gaskets, 33
Stony Brook (SUNY)
    radiative heaters, 8
Strain ellipsoid
    computer program, 92
    sample calculation, 101
    unit-cell parameters, 80–81
Suppliers
    names and addresses, 90–91
Symmetry
    phase transformation, 198
    polyhedral tilting, 199

Talc-like layer
    definition, 203
Temperature
    effect on bond distance, 121, 124, 133, 136
    effect on compressibility, 187–189
    effect on crystal energy, 119–120
    effect on crystal structure, 123, 137–138
    effect on ionic bond, 116
    effect on thermal expansion, 121, 187
    effect on thermal parameters, 139–142
    inverse relationship with pressure, 180
    structural analogy with P and X, 178

Temperature controller
    circuit diagram, 11
Tetrahedral rotation
    definition, 204
    illustration, 205
Thermal corrections
    bond distances, 82–83
Thermal expansion
    automated measurement of, 12
    effect of pressure on, 187–189
    effect of temperature on, 187
    polyhedra, 85, 180–181
Thermal parameters
    anharmonic vibrations, 87–88
    harmonic vibrations, 86–87
    least-squares refinement, 88
    models, 86
    pressure variation, 161
    rigid-body motion, 87
Thermal shielding
    Kapton, 9
Thermocouple
    contamination, 12
    high-temperature research, 14
    platinum, 12, 15
    radiative heaters, 9
Topology
    phase transformation, 194–195, 197
Transmission electron microscopy
    serpentine, 209
Twinning
    phase transformation, 200–201
    polyhedral tilting, 200–201

Unit-cell parameters
    determination, 50
    least-squares refinement, 50
    strain ellipsoid, 80–81
    variations, 79
    volume changes, 79

Vegard's law
    definition, 168
    micas, 168, 170
    oxides, 168, 169
Vibrations
    anharmonic, 87
    harmonic, 86–87
    molecular, 87
Volume
    bulk modulus relationship, 151–159
    calculation for polyhedra, 108
    unit-cell changes, 79–80

Volume of mixing
    excess function, 171–172

Weissenberg geometry
    heaters, 5

X-radiation
    molybdenum, 41
    silver, 42
    wavelengths, 42

Formula Index

Ag, 14
AgBr, 14
AgI, 14, 153
AlAs, 127
Al_2O_3, 122, 123, 126, 150, 152
Al_2SiO_5, 122, 131, 212
Ar, 29, 38, 40
Au, 14

BN, 127, 153
$BaBiO_3$, 132
BaF_2, 153, 154
$BaFeSi_4O_{10}$, 23, 44–45, 151, 197, 198, 199, 201
BaO, 125, 152, 168, 169
BaS, 153
BaSe, 153, 154
BaTe, 153, 154
$BaTiO_3$, 197
BeO, 126, 152, 154
Bi_2O_3, 126
Bi_2UO_6, 132
Br, 21

C, 127, 153
CCl_4, 21
CH_4, 29, 38
C_6H_6, 21
C_6H_5Cl, 59
CS_2, 21
$CaAl_2SiO_6$, 174
$CaAl_2Si_2O_8$, 131, 189
$Ca_3Al_2Si_3O_{12}$, 128, 151, 152, 185
$CaCO_3$, 19, 23, 70–72, 150, 186, 196, 197
CaF_2, 69–71, 73, 75, 123, 127, 149, 153, 154, 182

$CaFeSi_2O_6$, 130, 140, 173
$Ca(Mg,Fe)Si_2O_6$, 173
$(Ca,Mg,Fe,Ti,Al)(Si,Al)O_3$, 151, 152
$CaMgSiO_4$, 129, 133, 141
$CaMgSi_2O_6$, 75, 130, 133, 140, 141, 151, 155, 173, 174, 185
$Ca_2Mg_5Si_8O_{22}(OH)_2$, 132, 133, 141
$CaMnSiO_4$, 129
CaO, 123, 125, 152, 154, 168, 169
CaS, 153
CaSe, 153, 154
$CaSiO_3$, 212
CaTe, 153, 154
$CaTiSiO_5$, 129
CdI_2, 139
CdO, 125, 168, 169
CdS, 153, 154
CdSe, 153, 154
CdTe, 153, 154
CeO_2, 126
CoO, 125, 152, 168, 169
$CoSiO_3$, 210
Co_2SiO_4, 212
Cr_2O_3, 126, 150, 152
Cs, 21
$CsAuCl_3$, 150
CsBr, 127, 153, 154
CsCl, 123, 149, 153, 158, 182, 188, 189, 190
CsI, 153, 154
$CsMg_3AlSi_3(OH)_2$, 171
Cu, 14
CuCl, 153

EuO, 168, 169

230

(Fe,Mg,Ca)SiO$_3$, 132
(Fe,Mg)O, 173, 179
(Fe,Mg)SiO$_3$, 129, 132
(Fe,Mg)$_2$SiO$_4$, 173, 179
(Fe,Mg)Ti$_2$O$_5$, 129
FeO, 29, 125, 152, 168, 169, 173, 179
Fe$_2$O$_3$, 126, 150, 152
FeS, 150
FeSiO$_3$, 129, 210, 212
Fe$_2$SiO$_4$, 128, 150, 173, 179, 183, 192, 212
Fe$_9$Si$_6$O$_{20}$(OH)$_5$, 198
FeTiO$_3$, 172

Ga, 21
GaAs, 127, 153
GaP, 153
GaSb, 153
GeO$_2$, 150, 152

H$_2$, 25, 38
H$_2$O, 21
He, 38
HfO$_2$, 126
HgTe, 153

InAs, 153
InP, 153
InSb, 153

KAlSi$_3$O$_8$, 180, 182, 183, 190, 198, 201, 202, 203, 211, 212
KBr, 127, 152, 154
KCN, 153
KCl, 14, 127, 152, 154
KCo$_3$AlSi$_3$O$_{10}$(OH)$_2$, 170, 206
KCu$_3$AlSi$_3$O$_{10}$(OH)$_2$, 170
KF, 152, 154
KFe$_3$AlSi$_3$O$_{10}$(OH)$_2$, 170, 206
KI, 127, 153, 154
KMg$_3$AlSi$_3$O$_{10}$(OH)$_2$, 131, 133, 139, 151, 152, 170, 171, 185, 203, 204, 206
KMg$_3$BSi$_3$O$_{10}$(OH)$_2$, 170
K(Mg,Fe)$_3$AlSi$_3$O$_{10}$(OH)$_2$, 203
KMg$_3$FeSi$_3$O$_{10}$(OH)$_2$, 170
KMg$_3$GaSi$_3$O$_{10}$(OH)$_2$, 170
KMn$_3$AlSi$_3$O$_{10}$(OH)$_2$, 206
KNO$_3$, 21
KNi$_3$AlSi$_3$O$_{10}$(OH)$_2$, 170, 206

LiAlSiO$_4$, 132
LiAlSi$_2$O$_6$, 130, 140
LiBr, 152, 154

LiCl, 152, 154
LiF, 126, 152, 154, 188, 189, 190
LiI, 152, 154

(Mg,Al,Fe)$_6$(Si,Al)$_4$O$_{10}$(OH)$_8$, 203
Mg$_3$Al$_2$Si$_3$O$_{12}$, 128, 151, 152, 212
Mg$_2$Al$_4$Si$_5$O$_{18}$·n(H$_2$O), 131, 133, 198
MgF$_2$, 153, 154
(Mg,Fe)O, 125, 173, 179
(Mg,Fe)SiO$_3$, 6
(Mg,Fe)$_2$SiO$_4$, 128, 173, 179, 213
(Mg,Fe)$_6$Si$_8$O$_{20}$(OH)$_4$, 203
MgO, 69, 118, 122, 123, 125, 128, 133, 141, 150, 152, 154, 155, 168, 169, 173, 179, 183, 190
Mg(OH)$_2$, 139
MgSiO$_3$, 132, 149, 151, 155, 198, 210, 212
Mg$_2$SiO$_4$, 12, 128, 133, 139, 141, 149, 150, 152, 155, 173, 179, 183, 192, 212, 213
Mg$_3$Si$_2$O$_5$(OH)$_4$, 207–210
Mg$_3$Si$_4$O$_{10}$(OH)$_2$, 203
Mg$_7$Si$_8$O$_{22}$(OH)$_2$, 132
MgTiO$_3$, 172
MnF$_2$, 43, 150, 153
MnO, 125, 152, 154, 168, 169
MnO$_2$, 152
MnS, 127
MnSiO$_3$, 182, 183, 185, 210

NH$_4$Mg$_3$AlSi$_3$O$_{10}$(OH)$_2$, 171
NaAlSiO$_4$, 198
NaAlSi$_2$O$_6$, 130, 140, 190, 212
NaAlSi$_3$O$_8$, 80, 82, 83, 101–102, 130, 131, 152, 180, 182, 183, 184, 189, 197, 198, 199, 201, 202, 203
NaAlSi$_2$O$_6$·H$_2$O, 198, 201–202, 204
Na$_2$Al$_2$Si$_3$O$_{10}$·2H$_2$O, 131
NaBr, 152, 154
NaCl, 14, 15, 69, 123, 127, 149, 150, 152, 154, 161, 182
NaCrSi$_2$O$_6$, 130, 140
NaF, 14, 126, 152, 154, 188, 189, 190
NaFeSi$_2$O$_6$, 130, 140
NaI, 14, 153, 154
(Na,K,Ca)AlSiO$_4$, 131
NaMg$_3$AlSi$_3$O$_{10}$(OH)$_2$, 171
NaNO$_3$, 158
NaNO$_2$, 158
NaNbO$_3$, 198
NbC, 127
Ne, 25, 29, 38, 40
NiO, 125, 152, 168, 169

Ni_2SiO_4, 129, 150, 152, 182, 183, 212

Pb, 14
PbF_2, 153, 154
PbS, 127, 153
PbSe, 127, 153
PbTe, 127, 153
$PbTiO_3$, 129

RbBr, 127, 152, 154
RbCl, 152, 154
RbF, 152
RbI, 153, 154
$RbMg_3AlSi_3O_{10}(OH)_2$, 171
ReO_3, 126
RuO_2, 150, 152

Sb, 150
Se, 150
SiO_2, 5, 118, 126, 128, 150, 152, 159,
    183–184, 192, 198, 199, 201, 211,
    212
SnO_2, 150, 152
SnS_2, 151, 153
SnTe, 153
SrF_2, 153
SrO, 125, 152, 154, 168, 169
SrS, 153

SrSe, 153, 154
SrTe, 153, 154

TaC, 127, 153
Te, 150
ThBr, 153
ThCl, 153
ThO_2, 126, 152
TiC, 127, 153
TiN, 127
TiO_2, 123, 128, 150, 152, 186
Ti_2O_3, 128

UC, 153
UN, 127
UO_2, 126, 152

V_2O_3, 128, 150, 152, 158

ZnO, 126, 152, 154, 158
ZnS, 123, 127, 149, 153, 154, 182
ZnSe, 127, 153, 154
$ZnSiO_3$, 212
ZnTe, 153, 154
ZrC, 127, 153
ZrO_2, 123, 126
$ZrSiO_4$, 108–111, 150, 152